Pascal's Triangle:

A Study in Combinations

Jason VanBilliard

Acknowledgements

I would like to thank Dr. Joseph Moser for introducing me to the elegance of combinatorial reasoning, Jan Abramowitz for his detailed critique of my writing, and my wife and children for their encouragement and sacrifice of time during the writing process. *Soli Deo gloria*

Foreword

You may enjoy this book if you are mathematically curious and have a reasonably sound high school mathematics background. There is no calculus or other higher mathematics needed to study the contents of this book. The challenge of this excursion is that it involves a way of thinking that is unfamiliar to many: combinatorial argument. Consequently, it is my hope that by reading this book you will learn something new about the broader world of mathematics.

This book first introduces the reader to many interesting patterns in Pascal's Triangle. These patterns serve as the gateway to thinking in terms of combinations. This brief introduction to Pascal's Triangle and combinations is not meant to replace a deep mathematical study into the richness of combinatorial proof and discrete mathematics. However, this short trek is a pleasant and rewarding journey for the mind. For a deeper study in combinatorial proof, I recommend *Proofs that Really Count: The Art of Combinatorial Proof* by Arthur T. Benjamin and Jennifer Quinn.

Table of Contents

Chapter 0

Introduction

The mathematician-philosopher Blaise Pascal (1623-1662) was not the first to study the triangle which bears his name. The study of the triangle crossed many cultural and chronological boundaries. It can be found in the work of Chinese mathematicians Yang Hui (late thirteenth century), Chu Shih-Chien (1303), and Chia Hsien (1050). Indian mathematician Bhaskara (1114-1185) calculated a significant portion of the triangle while it was applied by Baghdad mathematicians al-Karaji and al-Samaw'al (1125-1180). European use predated Pascal as seen in works by Stifel, Tartaglia, Cardano, Viete, and Mersenne among others. Although many mathematicians had used and studied the triangle, it was Blaise Pascal who was the first to publish a comprehensive treatise.

Written in 1653, Pascal's *Traite du triangle arithmetique* was a culmination of correspondences between mathematician great Pierre de Fermat and Pascal regarding odds in games of chance. The perpendicular format in which he published the triangle is not the more familiar version that will be referred to in this study. Pascal portrayed his triangle as one of many with a similar structure. His particular triangle has a generator value of one.

Pascal's Triangle

With the generator in the first position (figure 0.1) each subsequent term is calculated:

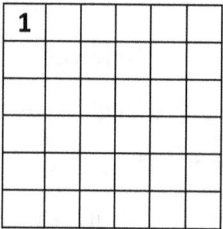

Figure 0.1

"The number of the first cell which is at the right angle is arbitrary; but once it has been placed all the other numbers are determined, and for this reason it is called the generator of the triangle. And every one of the other numbers is specified by this sole rule: The number of each cell is equal to that of the cell preceding it in its perpendicular rank plus that of the cell which precedes it in its parallel rank."

For example, in figure 0.2, the "15" is found by adding the "10" immediately preceding it to the left and the "5" above: 5 + 10 = 15. This pattern follows as shown below in figure 0.3. Pascal details to 10 rows and columns.

```
1  1  1   1   1   1
1  2  3   4   5   6
1  3  6  10  15
1  4  10 20
1  5  15
1  6
```

Figure 0.2

```
1   1   1   1   1   1   1   1   1   1
1   2   3   4   5   6   7   8   9
1   3   6  10  15  21  28  36
1   4  10  20  35  56  84
1   5  15  35  70 126
1   6  21  56 126
1   7  28  84
1   8  36
1   9
1
```

Figure 0.3

It should be noted that most modern presentations of Pascal's Triangle, including the use of the triangle in this book, rotate the triangle by 45 degrees (figure 0.4). Consequently, after constructing the diagonals of ones, each term is the sum of the two above it. When Pascal's Triangle is viewed this way, there are some indexing nuances. First, the top row with the single "1," is referred to as row "0" or the 0th row and the row containing "1, 3, 3, 1" is the 3rd row. Also the first number listed in each row is the 0th term. Therefore, the third row contains the 0th, 1st, 2nd, and 3rd terms. The value "6" in the figure below is in the 4th row, 2nd position.

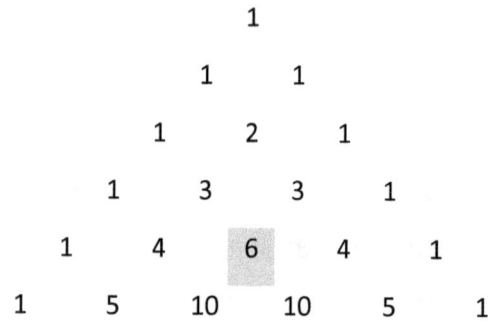

Figure 0.4

In part 1 of Pascal's treatment of the "Arithmetic Triangle," he identifies nineteen "consequences," or what we might call patterns, in the triangle. These observations include the "hockey stick pattern" and "powers of 2 pattern" that will be examined later. Part 2, the balance and bulk of his treatise, develops the application of the triangle in four areas: figurate-triangular type numbers, combinations, gambling stakes for games involving two players, and algebraic expansion of binomials. All four of these areas will also be examined in this book.

It will be valuable to familiarize yourself more with the structure and values in Pascal's Triangle as these values and the "sum of the two numbers above" pattern will be referred to in the following pages. Take some time to write Pascal's Triangle for yourself up to the eighth line and keep it at hand while reading.

Thinking with Combinations: Combinatorial Thinking

Despite the liberal use of Pascal's Triangle in this book, it is not primarily about Pascal's Triangle. Instead, the triangle provides a context for a studying an often neglected type of mathematical thinking. With some effort, many of us learned some algebraic rules, a handful of geometric properties, and fundamental logic in school. If we were fortunate to have had great teachers or insightful enough to develop our own deeper understanding, we may have truly developed algebraic thinking, proof thinking, and geometric thinking. Some mathematics programs even help students develop probabilistic and statistical thinking. However, most did not spend any significant time developing *combinatorial thinking* is school.

Combinatorics is at its most basic level the art of counting. It is simple, yet requires a different kind of mathematical ingenuity. In their book that reveals the unexpected development of combinatorial thinking by the Greek mathematician and scientist Archimedes, Netz and Noel (2007) describe combinatorial thinking as "in some ways, indeed, a simple science" which "can be approached without any complicated tools." Yet, they go on to say that, "it is almost as if for each new problem we need to invent a new, ingenious approach. Combinatorics is a science of endless ingenuity, of endless puzzles and games." (244-245). In this book you will experience this "simple science" that on the surface asks easily understood questions, yet requires your mind to develop in an alternate way.

Most modern middle and high school curricula have some sections devoted to combinations and permutations. Typical exercises in those sections may ask, "How many different committees of five can be made from among twelve people?" or "In how many different ways can 10 people form a line?" You may remember trying to recall the correct formula to apply in the given situation. Perhaps you did not experience these types of questions at all. In either case, most people do not get the opportunity to develop these very basic types of *combinatorial thinking* any further.

Our journey requires some healthy mental exercise. However, it is a gradual journey that produces better mathematical thinking. As you read, take time to reflect before moving on to the next portion. Try to predict results and develop your own arguments before reading those presented.

In this book, you will first be introduced to some of the many interesting patterns in Pascal's triangle. The patterns are inherently interesting to anyone casually interested in mathematics. After a brief introduction or re-introduction to combinations, Pascal's Triangle will be re-expressed in terms of combinations. You will then begin reexamining the patterns in Pascal's Triangle from a more intuitive, rewarding combinatorial perspective. Let's begin the journey.

Chapter 1

A Few Patterns in Pascal's Triangle

As Pascal indicated in his original treatise, his triangle is brimming with interesting properties or patterns. These patterns are numerous and extend across many different mathematical disciplines. Consequently, Pascal's Triangle may seem quite mysterious. The patterns can be applied in arithmetic, algebra, probability, basic number theory, and figurate numbers, and even help to gain access to the transcendental number e.

Binomial Theorem

The binomial theorem is probably the coolest tool you have forgotten from algebra. It is the ultimate shortcut in certain situations. You may recall the term "binomial" from some of the directions in a textbook: "Factor into two binomials" or "Multiply the monomial and the binomial." A binomial is simply the sum of two terms. For sake of ease, consider the binomial (x+y). Notice the coefficient of x is 1 and the coefficient of y is 1 (noted below as 1,1).

When we square the binomial $(x+y)^2=(x+y)(x+y)$ the result after distributing the multiplication is $x^2+xy+xy+y^2$. After simplification, the product is

$$1x^2+2xy+1y^2.$$

Notice the coefficients: <u>1,2,1</u>.

If you were to continue and multiply

$(x+y)^3=(x+y)(x+y)(x+y)$, you obtain

$$1x^3+3x^2y+3xy^2+1y^3.$$

Note the coefficients: <u>1,3,3,1</u>.

Similarly,

$$(x+y)^4 =1\, x^4 + 4x^3y + 6x^2y^2 + 4xy^3 +1\, y^4.$$

<u>(1, 4, 6, 4, 1)</u>.

It is important to remember for this discussion that $(x+y)^0=\underline{1}$.

As we examine the coefficients of the progressing binomial multiplication, a clear pattern emerges:

1

1,1

1,2,1

1,3,3,1

1,4,6,4,1

Reorganizing the coefficients into a more typical format:

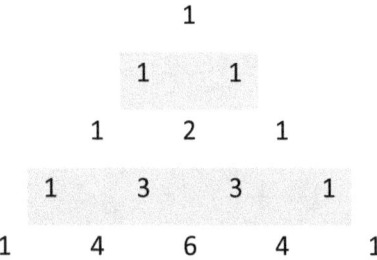

The pattern of coefficients corresponding directly to the values in Pascal's Triangle continues for all whole number values n in $(x+y)^n$. (Note that it is also easy to generalize this for the products of all binomials like $(3m+4nk)^4$ by simply substituting 3m for x and 4nk for y) This is a pretty cool pattern and makes algebra easy, but the rationale for its relationship to Pascal's Triangle is most easily understood as a consequence of combinatorial thinking.

Symmetry

To many the most obvious property of Pascal's Triangle is the line of symmetry running down the triangle (figure 1.1). The right and left sides mirror one another. The rules for constructing the triangle make symmetry necessary. However, there is also valuable combinatorial insight to be gained from even this basic observation.

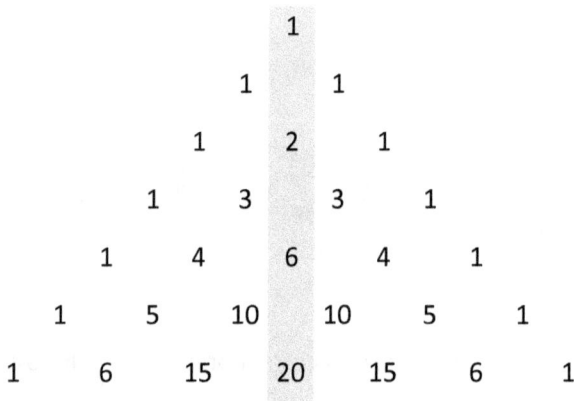

Figure 1.1

Sum of Each Row: 2^n

It is easy to miss a simple pattern in each row of Pascal's Triangle. If you add the rows going across, the 0th row has a sum of 1, the 1st row has a sum of 1+1= 2, the 2nd a sum of 1+2+1= 4, the 3rd a sum of 1+3+3+1= 8, and so on. A chart reveals the simple doubling pattern. Each sum represents a power of 2:

Row	Terms in Series	Sum	Power
0	1	1	2^0
1	1+1	2	2^1
2	1+2+1	4	2^2
3	1+3+3+1	8	2^3
4	1+4+6+4+1	16	2^4
5	1+5+10+10+5+1	32	2^5
N			2^n

As we will see later, understanding combinations provides insight as to why this pattern should not be surprising or mysterious, but should be expected in Pascal's Triangle.

Pascal's Hockey Stick

Continuing with the theme of adding terms, there is another interesting addition pattern. Most frequently it listed as the "Hockey Stick" pattern due to its visual representation. Choose any "Hockey Stick" in figure 1.2. Add the numbers going down the "shaft" of the stick and it equals the number on the "blade" of the stick. For instance, one hockey stick in figure 1.2 shows the sum 1+4+10+20 = 35.

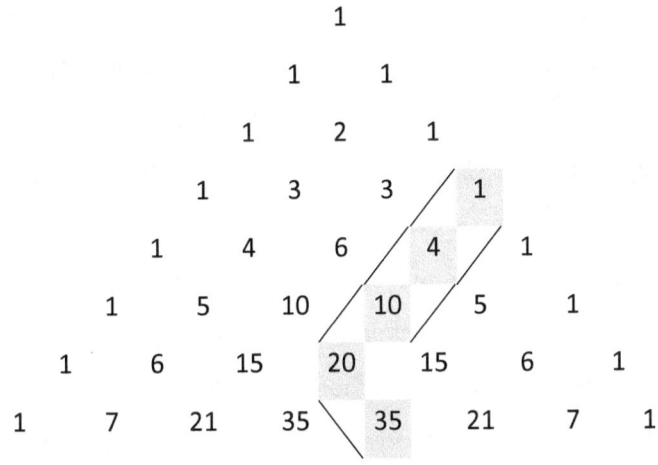

Figure 1.2

How do combinations make this an expected pattern in the triangle?

Fibonacci Numbers

Many math students are familiar with the Fibonacci Sequence. It derives its name from the 15th century Italian mathematician Leonardo of Pisa, also known as Fibonacci. He posed the following problem:

"How many pairs of rabbits can be bred from one pair in one year? A man has one pair of rabbits at a certain place entirely surrounded by a wall. We wish to know how many pairs can bred from it in one year, if the nature of these rabbits is such that they breed every month one pair and begin to breed in the second month after their birth."

The solutions to this problem reveal the following sequence: 1, 1, 2, 3, 5, 8, 13, 21.... This sequence is recursive meaning that each successive number is in some way derived from previous ones. For instance, consider a different sequence 1, 2, 4, 8, 16, 32, 64... . This simple sequence expressed recursively starts with 1. Each later term is two times the previous term: $S_0 = 1$ and $S_n = 2 \times S_{n-1}$, . (It can also be expressed explicitly: $S_n = 2^n$). The Fibonacci sequence does have an explicit formula, but is understood more easily recursively: start with 1 and 1 then each subsequent term is the sum of the two previous terms. For example 13 = 8+5 and 21 = 13+8. More generally: $F_0=1$, $F_1=1$, and $F_n = F_{n-1} + F_{n-2}$.

Where is the Fibonacci Sequence in Pascal's Triangle? Consider the sums of the numbers on the diagonals in figure 1.3 and the chart below.

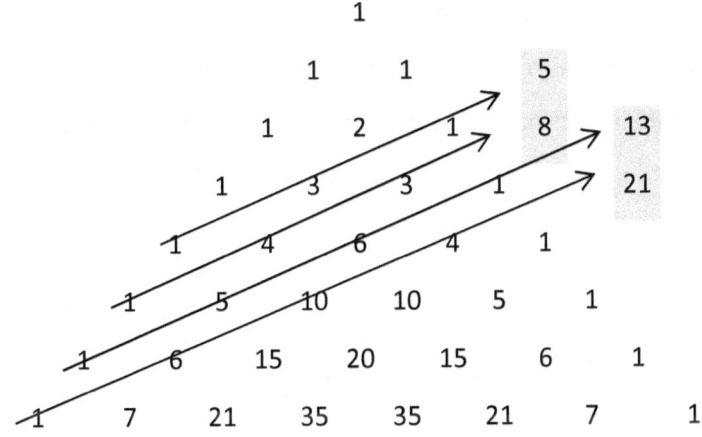

Figure 1.3

1	1
1	1
1+1	2
1+2	3
1+3+1	5
1+4+3	8
1+5+6+1	13

This may seem a curious and surprising pattern to find in Pascal's Triangle. However, an understanding of combinations convinces us why it should be there.

Powers of 11

If you had memorized the powers of 11, you would have immediately recognized another pattern in Pascal's Triangle. However, neither I nor anyone I know has committed them to memory without knowing the following facts. If you can generate Pascal's Triangle then you do not need to memorize the powers of 11. The pattern is easy at the beginning but is more complicated as one proceeds. It may be surprising that the pattern appears. However, as you will read later, combinations help us to realize that the patter should be there. Examine a list of the first few powers of 11 and look for the pattern in Pascal's Triangle:

11^0	1
11^1	11
11^2	121
11^3	1331
11^4	14641
11^5	161051
11^6	1771561

The pattern is obviously in Pascal's Triangle for rows 0-4, but the pattern seems to break down in row 5. Actually it does not. A brief examination of our base 10 number system helps to us to see the pattern's continuation. Let us consider $11^3 = 1331$. What does 1331 mean? It means adding one "10^3's" plus three "10^2's" plus three "10^1's" plus one "10^0" for a total of 1x1000 + 3x100 + 3x10 + 1x1, 1331. Similarly, $11^4 = 14641$ means you are adding one "10^4's" plus four "10^3's" plus six "10^2's" plus four "10^1's" plus one "10^0" for a total of 1x10000 + 4x1000 + 6x100 + 4x10 + 1x1, 14641. This pattern holds for 11^5. However, it gets a little more sophisticated.

Consider the table below:

Power of 11	Number of 1,000,000's	Number of 100,000's (10^5)	Number of 10,000's (10^4)	Number of 1,000's (10^3)	Number of 100's (10^2)	Number of 10's	Number of 1's	Total adding across
11^0							1	1
11^1						1	1	11
11^2					1	2	1	121
11^3				1	3	3	1	1,331
11^4			1	4	6	4	1	14,641
11^5		1	5	10	**10**	5	1	161,051
11^6	1	6	15	20	15	6	1	1,771,561

Pascal's Triangle is in the chart above. In Pascal's row 5, the boldface 10 does not seem to occur in 161,051. However, understanding the terms in the rows of Pascal's Triangle as representing the number of powers of 10 in the total, reveals the powers of 11. Completing the logical progression, 11^5 = one 100,000 plus five 10,000's plus ten 1,000's plus ten 100's plus 5 10's plus one 1 which adds to 161,051.

One 10^5	1 x 100000 =	100,000
Five 10^4	5 x 10000 =	50,000
Ten 10^3	10 x 1000 =	10,000
Ten 10^2	10 x 100 =	1,000
Five 10^1	5 x 10 =	50
One 10^0	1 x 1 =	1
	TOTAL	161,051

A power of 11 is in every row, it just gets hidden by our base 10 number system. Understanding why this pattern should be expected in Pascal's Triangle will be addressed later.

Figurate Numbers

Ancient Greek mathematicians investigate in what we call "Figurate Numbers." One , four, nine, and sixteen are called square numbers because one, four, nine, or sixteen stones, can be arranged into squares as seen below. [figure 1.4]

1	4	9	16
O	O O	O O O	O O O O
	O O	O O O	O O O O
		O O O	O O O O
			O O O O

Figure 1.4

Similarly, "triangular numbers" denote the number of stones that can be arranged into triangular shapes as illustrated below. [figure 1.5]

1	3	6	10
O	O	O	O
	O O	O O	O O
		O O O	O O O
			O O O O

Figure 1.5

The list of triangular numbers, 1, 3, 6, 10, 15..., can be expressed explicitly by the formula $T_n = \frac{n(n+1)}{2}$. This sequence can also be recursively understood as the nth triangular number is the previous triangular number plus n: $T_n = T_{n-1} + n$. The triangular numbers are easy to find in Pascal's Triangle in the third diagonal. (figure 1.6.)

Figure 1.6

The three dimensional analogy to the triangular numbers is known as the set of tetrahedral numbers. These figurate numbers are derived from the number of marbles needed to build tetrahedrons.

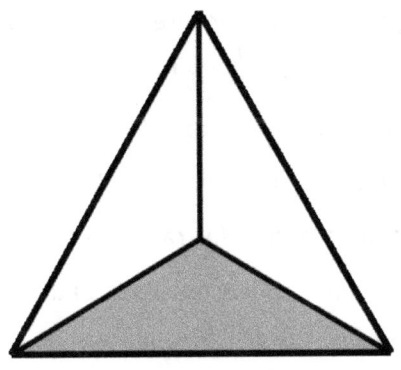

Tetradedron

The tetrahedral numbers, 1, 4, 10, 20, 35..., can be seen in the fourth diagonal of Pascal's Triangle (figure 1.7)

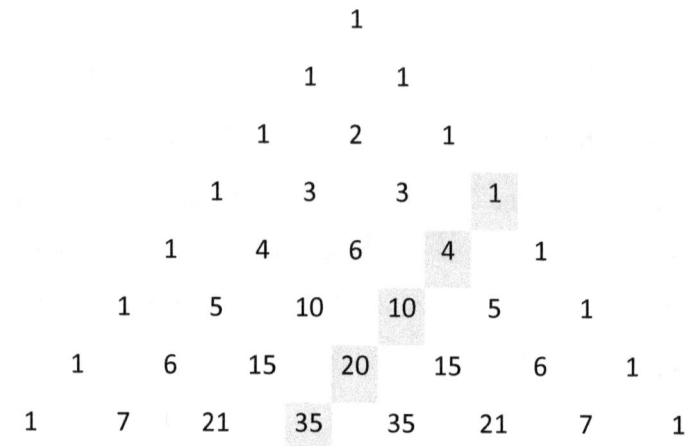

Figure 1.7

For this pattern to continue we will have to exit 2- and 3-dimensions and drop to 1- or 0-dimensions or up to 4 or more dimensions. The pattern would go something like this: the two-dimensional triangular numbers are on the third diagonal while the three-dimensional triangular-like numbers (the tetrahedral numbers) are on the fourth diagonal. Then the four-dimensional triangular-like numbers are on the fifth diagonal and so on. I am not sure they can be easily illustrated on this two-dimensional page you are reading. However, their patterns can be easily extrapolated and are clearly in Pascal's Triangle.

Why should these patterns appear in Pascal's Triangle? An understanding of the construction of combinations will reveal the reason.

Binomial Probabilities

Binomial probabilities address either/or, on/off, up/down, or any other type of 0/1 situation. A familiar version of this situation involves tossing a coin. The two options are heads (H) and tails (T). The relationship between binomial probability situations and Pascal's Triangle can easily be extended to scenarios that do not involve a 50% - 50% probability. However, here we will limit ourselves to the basic 50-50 construct.

Suppose you flip one coin. The possible outcomes are H or T. Flipping a coin two times results in four possible outcomes: first flip H, second flip H (HH) OR first flip H, second flip T (HT) OR first flip T, second flip H (TH) OR first flip T, second flip T (TT). So the four possible outcomes are HH, HT, TH, and TT. To consider the situation another way, there is a ¼ chance of getting 0 tails, a 2/4 chance of getting 1 tails, and a ¼ chance of getting 2 tails. The probabilities are $\frac{1}{4}, \frac{2}{4}, \frac{1}{4}$. It is no coincidence that the numerators 1, 2, 1 appear in the second row of Pascal's Triangle.

Suppose you flip a coin three times. The possible outcomes are HHH, HHT, HTH, HTT, THH, THT, TTH, and TTT. The probability of getting 0 tails: $\frac{1}{8}$; One tail: $\frac{3}{8}$; Two tails: $\frac{3}{8}$; Three tails: $\frac{1}{8}$. The sequence of numerators is the familiar 1, 3, 3, 1 of Pascal Triangle's third row.

This pattern continues. The highlighted "15" below indicates the number of ways to get exactly 4 tails when a coin is flipped 6 times.

Number of ways to get N tails → Number of flips ↓	0	1	2	3	4	5	6
0	1						
1	1	1					
2	1	2	1				
3	1	3	3	1			
4	1	4	6	4	1		
5	1	5	10	10	5	1	
6	1	6	15	20	15	6	1

Why should this pattern appear in Pascal's Triangle? The answer lies in combinatorial argument.

Chapter 2

Combinations

Combinations are the basis for the reasoning that will be developed in this book and are inherently related to Pascal's Triangle. A combination in mathematics is similar to other instances of "combinations" with which you may be familiar. Your high school locker may have had a 3-digit combination. It was a collection of 3 numbers usually from the numbers 1-40 that you used in a particular order. However, the mathematical concept of combination diverges a little from your locker combination. In mathematics, the combination 18-34-12 is the same as the combination 12-18-34. On your lock the order of the numbers matters; in mathematical combinations, order does not matter.

The situations in which order matters are called permutations in mathematics. One other important note is that the "combination" 18-34-18 does not qualify as a mathematical combination where a particular number cannot be repeated. Given a set of numbers or objects of size n, the total number of different combinations of size r that can be formed from the set of size n is denoted by the function $C(n,r)$. An alternative notation to $C(n,r)$ that will be used throughout this book is $C(n,r) = \binom{n}{r}$.

A few examples are instructive. Suppose you have 6 objects labeled [A,B,C,D,E,F]. How many ways can they chosen 2 at a time? We can easily list the 15 possibilities: AB, AC, AD, AE, AF, BC, BD, BE, BF, CD, CD, CF, DE, DF, EF. Recall that AB is the same as BA in mathematical combinations so AB and BA should only be counted as 1.

Suppose there a 5 people, A, B, C, D, and E, being considered for a 3 person committee. The total number of committees that could be formed (remember that in a committee, the order of names does not matter) is $\binom{5}{3}$ = 10: ABC, ABD, ABE, ACD, ACE, ADE, BCD, BCE, BDE, and CDE. If instead of 5 people being considered for a 3 person committee there were 7 people being considered, the total number of committees (combinations) is $\binom{7}{3}$ = 35.

All three of these numbers ($\binom{6}{2}$ = 15, $\binom{5}{3}$ = 10, and $\binom{7}{3}$ =35) are found in Pascal's Triangle. Given the convention that the top of Pascal's Triangle is row 0 and the first position in each row is position 0, the combination notation indicates exactly where the number is found. For instance, $\binom{6}{2}$ = 15 is found in row 6 position 2. Similarly, $\binom{5}{3}$ = 10 is found in row 5 position 3

and $\binom{7}{3}$ =35 is found in row 7 position 3. $\binom{7}{3}$ could be

substituted for 35 in row 7 position 3. Hence, Pascal's Triangle

could be rewritten as in figure 2.1 below.

This relationship between combinations and Pascal's Triangle is

due to the inherent nature of each. Recall that Pascal's Triangle

is built by starting with a generator of 1 and then "adding the

two above" to construct triangle. Combinations also have an

"adding the two above" property which will be demonstrated

later. Consequently, an understanding of combinations

demystifies many amazing patterns in Pascal's Triangle and

develops the mind of the one who studies both.

$$\binom{0}{0}$$

$$\binom{1}{0}\quad\binom{1}{1}$$

$$\binom{2}{0}\quad\binom{2}{1}\quad\binom{2}{2}$$

$$\binom{3}{0}\quad\binom{3}{1}\quad\binom{3}{2}\quad\binom{3}{3}$$

$$\binom{4}{0}\quad\binom{4}{1}\quad\binom{4}{2}\quad\binom{4}{3}\quad\binom{4}{4}$$

$$\binom{5}{0}\quad\binom{5}{1}\quad\binom{5}{2}\quad\binom{5}{3}\quad\binom{5}{4}\quad\binom{5}{5}$$

$$\binom{6}{0}\quad\binom{6}{1}\quad\binom{6}{2}\quad\binom{6}{3}\quad\binom{6}{4}\quad\binom{6}{5}\quad\binom{6}{6}$$

$$\binom{7}{0}\quad\binom{7}{1}\quad\binom{7}{2}\quad\binom{7}{3}\quad\binom{7}{4}\quad\binom{7}{5}\quad\binom{7}{6}\quad\binom{7}{7}$$

Figure 2.1

In order to fully develop thinking in terms of combinations, it is worth taking a few moments to investigate the locker type "combinations" in which repeating is allowed that are eliminated when focusing on mathematical combinations. There are two criteria that must be considered when counting arrangements of items: (1) does order matter? And (2) is there "replacement" – that is, can an element be repeated? These considerations produce a 2x2 chart:

Counting arrangements of n items taken r at a time.	Order matters (AB is different than BA)	Order does NOT matter (AB is the same as BA)
No replacement/ NO repeating of elements	1.	2.
Replacement/ repeating of elements	3.	4.

Mathematical combinations meet the requirements for box number " 2." and is denoted by nCr or C(n,r) or the notation used in this text: $\binom{n}{r}$. Box number "1" describes what are called permutations and is denoted nPr. Box number "3" is simply derived directly from the multiplication counting principle below. Box "4" is a bit more complicated and will be discussed in the last chapter as $\binom{n + r - 1}{n - 1}$, when the reader

has developed a deeper understanding of combinations.

Box "3" involves choosing r items from a collection of size n where order matters but the n items are always available even after each choice is made. An practical parallel situation would involve making "words" from our 26 letter alphabet. Suppose we wanted to make 3 letter "words" where a "word" is any 3-letter arrangement. "ABA" and "CFG" would examples of words. Furthermore, "ABA" is a different word from "AAB." The way to find the total number of "words" would be to consider the number of options in each position and multiply them: 26 options for the first position x 26 options for the second position x 26 options for the third position. (Multiplication should be the obvious operation: for example, an "alphabet" of 3 letters (A,B,C) and "words" of length 2. There are "3x3" words: A first then matched with each of the 3 and B first then matched with each of the 3 and C first then matched with each of the 3. The total is the result of repeated addition: multiplication. In this case there are 3 groups of 3.) The total number of 3 letter "words" from the 26 letter alphabet is 26 x 26 x 26 or 26^3. Similarly, the number of 5 letter "words" is 26^5. If we used the 24 letter Greek alphabet then the number of 5 letter "words" is 24^5. One can easily generalize this result of arranging n items taken r at a time where order matters and replacement is allowed as n^r.

Box "1" permutations lead directly to the focus of this book: combinations. Consider box "1" which involves the number of ways to arrange r out of n items when there is NO replacement and order DOES matter. The argument is similar to the argument for box "3" except that letters can NOT be repeated. Consequently "ABA" is not a "word" in this case. How many 3 letter "words" are there when letters can NOT be repeated? Try to answer this question before moving on.

The number of 3 letter "words" if letters can NOT be repeated can be calculated using the same principle as above. There are 26 options for the first letter. However, once one is used there are 25 options for the second letter then 24 options for the third letter. Therefore, there are 26 x 25 x 24 possible 3 letter "words" under these conditions. The number of 5 letter words would be 26 x 25 x 24 x 23 x 22. The number of 5 letter Greek words would be 24 x 23 x 22 x 21 x 20.

Generalizing this result takes a bit of algebraic manipulation but it follows directly from the examples above. For sake of simplicity, consider a 10 letter alphabet. How many 3 letter "words" are able to be generated in the" order matters" but "no repeating" conditions? Obviously there are 10 x 9 x 8 "words." Generalizing this situation for an n-lettered alphabet and r-letter words is simply n x (n-1) x (n-2) x ... x (n-r+1). However, a little arithmetic produces a simpler formula. Consider that

$$10 \times 9 \times 8 = \frac{10 \times 9 \times 8 \times 7 \times 6 \times 5 \times 4 \times 3 \times 2 \times 1}{7 \times 6 \times 5 \times 4 \times 3 \times 2 \times 1}$$

since the ratio 7x6x5x4x3x2x1 in the numerator and denominator is 1. In mathematics there is a operation called factorial that is used as a shortcut to writing a product like 7x6x5x4x3x2x1. Factorial notation employs an exclamation mark: 7! = 7x6x5x4x3x2x1. Consequently,

$$10 \times 9 \times 8 = \frac{10 \times 9 \times 8 \times 7 \times 6 \times 5 \times 4 \times 3 \times 2 \times 1}{7 \times 6 \times 5 \times 4 \times 3 \times 2 \times 1} = \frac{10!}{7!}$$

This factorial shortcut allows us to generalize the permutations in box "1." To summarize, n items taken r at a time when order matters and there is NO replacement is $nPr = \frac{n!}{(n-r)!}$.

The box "2" combinations that are the focus of this book can be derived from the permutations formula with one minor modification. The difference between combinations and permutations is order. In combinations order does NOT matter. In permutations order does matter. In both cases there is NO replacement. So the question to consider is "how many combinations are there of n items taken r at a time when order does NOT matter and there is NO replacement?"

Suppose we had a formula for box "2" = nCr = C(n,r) = $\binom{n}{r}$ often said as "n choose r." How would it relate to nPr? Consider an example. Suppose we are trying to make 3 letter "words" from a 5 letter alphabet [A,B,C,D,E]. In permutation the "word" ABC is different from ACB, BAC, BCA, CAB, and CBA. However, in combinations those 6 "words" are considered the same. Consequently, for *every* combination "word" there are 6 permutation " words." Or, as a formula:

$$_5P_3 = \binom{5}{3} \times 6$$

Where did the "6" come from? For nCr we would choose 3 letters, then order them to find the permutations of that choice. Three distinct letters can be organized in 3x2x1 = 3! = 6 different ways. To generalize, if we would choose r letters and order them, the number of orders would be r! for each choice. For the specific example:

$$_5P_3 = \binom{5}{3} \times 3!$$

More generally:

$$nPr = \binom{n}{r} \times r!$$

Which algebraically becomes

$$\binom{n}{r} = \frac{nPr}{r!} = \frac{\frac{n!}{(n-r)!}}{r!} = \frac{n!}{(n-r)!r!}.$$

A good way to understand box "4" is to consider an analogy: an ice cream store that serves bowls of ice cream. In this ice cream store there are n flavors that never run out! A person can come in and order a bowl of ice cream cone of r scoops. Suppose the store has 3 flavors. How many different 4 scoop bowls could be ordered if order does not matter and flavors can be reused. The answer is $\binom{3+4-1}{2}$. The rationale for this conclusion will be developed in the last chapter once the reader has a deeper understanding of combinations.

As a result of our exploration, we can fill in the chart as follows:

Counting arrangements of n items taken r at a time.	Order matters (AB is different than BA)	Order does NOT matter (AB is the same as BA)
No replacement/ NO repeating of elements	1. Permutations: $nPr = \dfrac{n!}{(n-r)!}$	2. Combinations: $nCr = C(n,r) = \binom{n}{r} = \dfrac{n!}{(n-r)!r!}.$
Replacement/ repeating of elements	3. n^r	4. $\binom{n+r-1}{n-1}$

Chapter 3

Algebraic Argument and

Combinatorial Argument

Although many have spent extensive hours in various algebra classes, few can claim to have had a course in combinatorics. Some have had a brief exposure to combinations in the rarely seen chapter 13 of an algebra 2 book or if the subtleties of a statistics were grounded in probability distributions. In any case, the American educational system has historically been strong in developing algebraic skills, some Euclidean geometric thinking and proof, and preparing students to move on to briefly grasp the holy calculus grail. However, the system has been deficient in deeply developing the minds of students in other exciting areas of mathematics such as number theory, probability, game theory, fractal geometry, topology, and combinatorics, to name a few.

Both the familiar and neglected areas of mathematics are accessible to everyone with some basic study but take significant effort to master. The goal of this book is to help you begin to develop one of these other areas of mathematical thinking. In this chapter you will see the difference between algebraic argument and combinatorial argument (an argument involving combinations). The point is not to downplay the power of algebraic argument but to open your mind to an alternate way of thinking about mathematics. Furthermore, in the following cases, I hope you find the combinatorial arguments in some way more satisfying than the algebraic arguments thereby motivating you to further study.

What makes one mathematical argument superior to another? Most people involved in mathematics have an intuitive sense of quality proof as there are very few objective criteria. However, there are two criteria worth considering. One criterion is the simplicity of the argument; does one argument take a more direct route than another? I hope that the simplicity of the combinatorial arguments involved with the patterns seen in Pascal's Triangle is in this way more satisfying than the algebraic arguments. However, seeing the simplicity may take a bit of expanding and adjusting your thinking strategies. Second, the proof argument is strong when it is clearly or easily meaningfully connected to the problem under consideration. Amidst an algebraic argument, the mathematician frequently is able to ignore the meaning of the problem under consideration. Instead she applies algebraic tools or rules that are powerful and secure while ignoring or veiling meaning. These properties will be demonstrated below.

Algebraic Arguments

Algebraic thinking is different than algebraic argument. Algebraic thinking is usually associated with the ability to recognize and extend or summarize pattern. Algebraic thinking is intentionally imbedded in quality K-12 curricula. For instance, in first grade my daughter frequently brought home math worksheets that have a specific section indicating "Algebraic Thinking." Obviously she is not sitting at home solving equations. However, she is asked to extend patterns then write a rule for the pattern she has extended. This "algebraic thinking" leads to the familiar linear, quadratic, exponential, and trigonometric equations of high school algebra.

When we use algebra rules associated with the high school courses to make argument in algebra or other areas of mathematics, we are employing algebraic argument. It is frequently used in combination with other ways of thinking but at times may stand on its own. Two examples should help to explain how algebraic argument is used in geometry and number theory.

The famous Pythagorean Theorem states that in any right triangle, the square of the hypotenuse equals the sum of the squares of the other two sides. Geometrically, what is happening is that literally the area of the square that is made by the length of the hypotenuse is the same as the total area of the two squares formed by the two sides. As seen in figure 3.1, the area of the lightly shaded square plus the area of the dark shaded square equals the area of the large square.

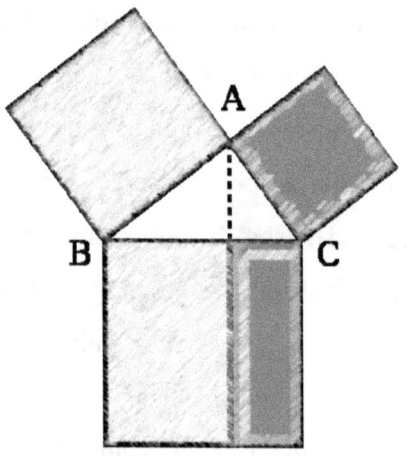

Figure 3.1

There are many geometric proofs of the theorem including the one found in Euclid's *Elements* written c. 300 B.C. However, it can also be proven with the support of algebra. In this case, the goal is to show algebraically that $a^2+b^2=c^2$ (a format which is more familiar to most). Here is one example of an algebraic argument.

Consider figure 3.2. The total area of the square can be calculated in two ways: (1) the area of the entire square using the outside dimensions or (2) adding the areas of the five figures inside the square.

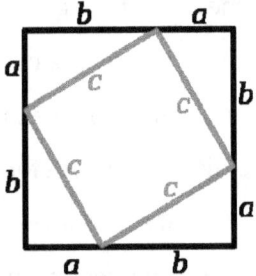

Figure 3.2

Taken the first way, the area of the square is (a+b)(a+b). Taken the second way, the area is the sum of four triangles plus the little square in the middle. Since the area of a triangle is ½ base x height, the area of each of the four triangles is ½ a x b. Finally the area of the little square is c x c = c^2. Both calculations are correct, therefore both must be equal.

Algebraically:

(1) (a+b)(a+b) = (½ a x b)+ (½ a x b)+ (½ a x b)+ (½ a x b)+ c^2

Using algebraic rules of expanding and simplifying:

(2) a^2 + ab + ab + b^2= 4(½ ab) + c^2

(3) a^2 + 2ab + b^2 = 2ab + c^2

Subtracting 2ab from each side of the equation:

(4) a^2 + b^2 = c^2

Notice that once the algebraic equation was set up, the algebraic argument proceeded without needing to understand the geometric meaning of the process involved. For instance, between step 2 and 3 we simplify 4(½ ab) to 2ab. Geometrically, we are maneuvering the 4 triangles into 2 rectangles. That geometric meaning or interpretation is not needed to proceed through the algebraic argument.

A second example may be helpful in understanding what is meant by algebraic argument before considering specific examples involving Pascal's Triangle and Combinatoric Argument. This example emerges for basic number theory. First, a few questions to consider: When one adds two even numbers, is the sum even or odd? When one adds two odd numbers, is the sum even or odd? When one multiplies two even numbers, is the product even or odd? When one multiplies two odd numbers, is the product even or odd? Each of these questions will be answered with an algebraic argument. The algebraic number theory argument, as with the Pythagorean Theorem argument, will begin with the nature of the problem, but then move to a purely algebraic approach before returning to the nature of the problem.

The sum of two evens. Every even number can be written as two times some integer (the set of integers: {...-3,-2,-1,0,1,2,3...}) (e.g. 120 = 2x60 and 38 = 2x19). So one can express two different even numbers as 2n and 2m where n and m are integers. Algebraically:

(1) $2n + 2m =$

(2) $2(n+m)$

n + m is some integer, so 2 (n+m) is an even number. Consequently the sum of two even numbers is always an even number.

The sum of two odds. Every odd number can be written an even number plus one (e.g. 121 = 120 + 1 = 2x60 + 1 and 39 = 38 + 1 = 2x19 + 1). So one can express two different odd numbers as 2n+1 and 2m+1 where n and m are integers. Algebraically:

(1)   (2n+1) + (2m+1) =

(2)   2n+2m+2 =

(3)   2(n+m+1)

n + m +1 is some integer, so 2 (n+m+1) is an even number. Consequently the sum of two odd numbers is always an even number.

The product of two evens. Let 2n and 2m where n and m are integers be two even numbers. Algebraically:

(1)   2n x 2m =

(2)   4nm=

(3)   2(2nm)

2nm is some integer, so 2(2nm) is an even number. Consequently the product of two even numbers is always an even number.

The product of two odds. Let 2n+1 and 2m+1 where n and m are integers be two odd numbers. Algebraically:

(1) $(2n+1)(2m+1) =$

(2) $4nm+2n+2m+1 =$

(3) $2(2nm+n+m)+1$

$2nm+n+m$ is some integer, so $2(2nm+n+m)$ is an even number so $2(2nm+n+m)+1$ is an odd number. Consequently the product of two odd numbers is always an odd number.

Algebraic Argument in Pascal's Triangle

As seen in the geometric and number theoretic examples above, algebraic arguments are powerful and secure, yet the use of algebraic argument allows the thinker to frequently leave the essence or meaning of the problem on the journey to the end. In this case, the ends do not justify the means; rather, the means justify the ends and have little apparent connection to the ends. Furthermore, they are not necessarily simple in the sense that they are not intuitively clear based on the problem presented. Instead, as powerful as algebraic argument is, it can frequently allow the thinker to leave the problem at hand and journey into a world of accepted rules before returning to the realities of the problem.

The two examples from patterns in Pascal's Triangle provided below contrast algebraic argument and combinatorial argument.   The simplicity and meaningful nature that combinatorial argument may have over algebraic argument are exemplified. A casual examination of Pascal's Triangle reveals two simple patterns. First, Pascal's Triangle is reflectively symmetric; the right side is the same as the left. Second, in Pascal's Triangle, each term is the sum of two terms "above" it. These two patterns will be the basis for the example below.

The other quality of Pascal's Triangle that will be considered in the arguments is that each of the terms represents a number of combinations as was illustrated in the previous chapter. The reader will recall that the 5th row, 2nd element (10) represents the number of ways of choosing pairs (2) of people from a group of five (5). For example, given people A,B,C,D, and E, the possible pairings are AB, AC, AD, AE, BC, BD, BE, CD, CE, and DE. This understanding that Pascal's Triangle contains combinations is the underpinning of the combinatorial thinking related to Pascal's Triangle. It is important that the reader understand that there is an algebraic formula to calculate combinations that was developed in chapter 2: the number of ways to choose r items from a list of n items where items cannot be reused and order does not matter is $\binom{n}{r} = \frac{n!}{(n-r)!r!}$ where "!" means factorial. (e.g. 5! = 5x4x3x2x1). As in the example above, choosing 2 people from a group of 5:

(1) $\qquad C(5,2) = \frac{5!}{(5-2)!2!} =$

(2) $\qquad \frac{5!}{(3!)!2!} =$

(3) $\dfrac{5\times4\times3\times2\times1}{(3\times2\times1)(2\times1)} =$

(4) $\dfrac{5\times4}{2\times1} =$

(5) $\dfrac{20}{2} =$

(6) 10.

Using the formula for combinations, it is fairly straightforward to show that the two patterns described above hold true: Pascal's Triangle is symmetric and each term is the sum of the two above. The algebraic arguments will be presented here followed by the corresponding combinatorial arguments. It is my hope that the combinatorial arguments will be a delightful taste of the kind of thinking that will begin to be developed through this book.

Algebraic Argument: Pascal's Symmetric Triangle

In order to convert the notion that Pascal's Triangle is symmetric into an algebraic argument, it is necessary to take view Pascal's Triangle in terms of combinations. Algebraically, $\binom{n}{r} = \dfrac{n!}{(n-r)!r!}$ represents the rth element in the nth row of Pascal's Triangle assuming counting starts at 0 as explained in chapter 1. Again, using combination notation, Pascal's Triangle can be rewritten as seen in figure 3.3.

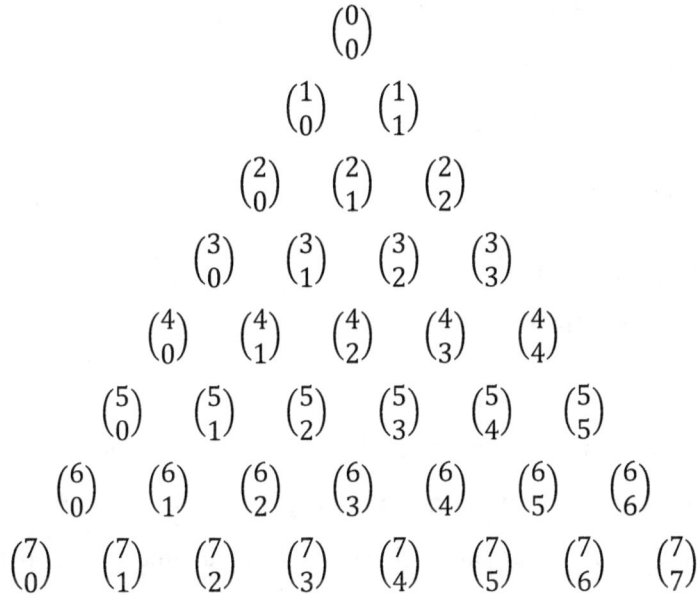

Figure 3.3

Examining any particular row will reveal the symmetry in a slightly different way. Consider row 6 and row 7. The following equalities occur:

$$\binom{6}{0}=\binom{6}{6}; \quad \binom{6}{1}=\binom{6}{5}; \quad \binom{6}{2}=\binom{6}{4};$$

$$\binom{7}{0}=\binom{7}{7}; \quad \binom{7}{1}=\binom{7}{6}; \quad \binom{7}{2}=\binom{7}{5}; \quad \binom{7}{3}=\binom{7}{4};$$

Based on these observations, in order to establish the symmetry of Pascal's Triangle everywhere one must prove that $\binom{n}{r}=\binom{n}{n-r}$. Based on the algebraic formula for combinations, this is a straightforward proof:

(1) $\quad \binom{n}{r} = \dfrac{n!}{(n-r)!r!} =$

Using a little algebra trick of adding nothing to a term (in this case adding 0 = n-n to r in the denominator) yields:

(2) $\quad \dfrac{n!}{(n-r)!(n-n+r)!} =$

Factoring out -1:

(3) $\quad \dfrac{n!}{(n-r)!(n-(n-r))!} =$

Using the commutative property:

(4) $\quad \dfrac{n!}{(n-(n-r))!(n-r)!} = \dbinom{n}{n-r}.$

This proof is fairly straightforward and required very little algebra, yet it is an algebraic argument for symmetry. The middle of the argument is simply algebraic manipulation that requires no understanding of Pascal's Triangle as containing combinations. Consider now the combinatorial argument.

Combinatorial Argument: Pascal's Symmetric Triangle

Because the reader has likely not thought combinatorially on a regular basis, it may be helpful to consider a concrete combinatorial example before arguing the general combinatorial case. Consider the following equality as shown above.

$$\binom{7}{2} = \binom{7}{5}$$

We would like to understand in a non-algebraic way why $\binom{7}{2} = \binom{7}{5}$. Alternately stated, we want to understand why 7 choose 2 is the same as 7 choose 5. Take a moment to try to argue it in your mind. The left side of the equation is looking for the number of ways to make groups of size 2 out of 7 elements. How many are there? How would the total be found? One could simply take the letters A, B, C, D, E, F, and G and exhaustively list the groups of size 2: AB, AC, AD...etc. The right side of the equation is looking for the number of ways to make groups of size 5 out of 7 elements. How many are there? How would the total be found? One could simply take the letters A, B, C, D, E, F, and G and exhaustively list the groups of size 5: ABCDE, ABCDF, ABCDG...etc. This would prove the symmetry for Pascal's Triangle for only one case. Is there something here to learn that can be generalize to all cases?

Consider some of the groups of size 5. When we chose to include ABCDE in a set which elements are left out? Elements F and G were left out. When we chose ABCDF the letters E and G are left out. Generalize this pattern. Choosing 5 elements from 7 elements to *be in* a group is the same as choosing 2 elements from 7 elements to be *left out* of a group. In other words, $\binom{7}{2}$ is the number of ways of choosing elements to *be in* a group of size 2 or alternately the number of elements to be *left out* of a group when 5 elements are chosen. Choosing 5 out of 7 to be in a group is *by default* choosing 2 out of 7 to be left out of the group. (This situation kind of reminds me of when we picked teams on the playground: Captain Brian not choosing me to be on his team could ultimately be seen as Brian choosing to place me on Captain Wendell's team.) To state it simply: choosing 5 people to be in is choosing 2 people to be out. Consequently it should be no surprise that $\binom{7}{2}=\binom{7}{5}$. This exemplifies combinatorial reasoning on a small scale.

It should now be easy to generate a combinatorial argument for the symmetry of Pascal's Triangle. To state the problem plainly again, it must be shown that $\binom{n}{r}=\binom{n}{n-r}$. The combinatorial argument is now straightforward.

Given n items, choosing r items to be in a set by default leaves n-r items out. Therefore every time a set of size r is made, a set of size n-r is also made. Consequently, the number of ways of choosing r items for a set is the same as the number of ways of choosing n-r for a set. Hence, $\binom{n}{r}=\binom{n}{n-r}$.

This argument uses no algebra. Instead, it engages the mind to think about what equality $\binom{n}{r}=\binom{n}{n-r}$ actually means. I would argue that this is a simpler, more intuitive way of looking at the symmetry of Pascal's Triangle. Consider now a second but more complex comparison.

Algebraic Argument: In Pascal's Triangle, Each Term is the Sum of the Two Above It

The algebraic argument for the "sum of the two above" pattern in Pascal's Triangle is also fairly straightforward. However, it strays even further from an intuitive understanding of why the identity works than the algebraic symmetry argument above. First, we must make a conjecture as to the general identity that needs to be proven. Using the version of Pascal's Triangle that involves combinations as seen in figure 3.4 below, the general pattern is fairly easy to define.

$$\binom{0}{0}$$

$$\binom{1}{0} \quad \binom{1}{1}$$

$$\binom{2}{0} \quad \binom{2}{1} \quad \binom{2}{2}$$

$$\binom{3}{0} \quad \binom{3}{1} \quad \binom{3}{2} \quad \binom{3}{3}$$

$$\binom{4}{0} \quad \binom{4}{1} \quad \binom{4}{2} \quad \binom{4}{3} \quad \binom{4}{4}$$

$$\binom{5}{0} \quad \binom{5}{1} \quad \binom{5}{2} \quad \binom{5}{3} \quad \binom{5}{4} \quad \binom{5}{5}$$

$$\binom{6}{0} \quad \binom{6}{1} \quad \binom{6}{2} \quad \binom{6}{3} \quad \binom{6}{4} \quad \binom{6}{5} \quad \binom{6}{6}$$

$$\binom{7}{0} \quad \binom{7}{1} \quad \binom{7}{2} \quad \binom{7}{3} \quad \binom{7}{4} \quad \binom{7}{5} \quad \binom{7}{6} \quad \binom{7}{7}$$

By examining a few examples the conjecture easily emerges. By inspection:

$$\binom{6}{4} = \binom{5}{4} + \binom{5}{3} \qquad \binom{7}{3} = \binom{6}{3} + \binom{6}{2}$$

$$\binom{3}{1} = \binom{2}{1} + \binom{2}{0} \qquad \binom{5}{4} = \binom{4}{4} + \binom{4}{3}$$

It is fairly easily generalized that what needs to be proven is that:

$$\binom{n}{r} = \binom{n-1}{r} + \binom{n-1}{r-1}$$

Algebraically this should not be too difficult by using the definition of combination. However, it is not as intuitively satisfying as the combinatorial argument. Furthermore, the combinatorial argument will be instructive in developing a new way of thinking. The most important piece to remember in the algebraic argument is the behavior of factorial (!). Recall that 5! = 5x4x3x2x1 so 5! could be rewritten as 5 x 4! = 5 x 4x3x2x1. In general terms, n! = n (n-1)!.

The Algebraic Argument that $\binom{n}{r} = \binom{n-1}{r} + \binom{n-1}{r-1}$.

Start with the right side of the identity:

(1) $\quad \binom{n-1}{r} + \binom{n-1}{r-1} = \dfrac{(n-1)!}{(n-1-r)!(r)!} + \dfrac{(n-1)!}{((n-1)-(r-1))!(r-1)!}$

(2) $\quad = \dfrac{(n-1)!}{(n-r-1)!(r)!} + \dfrac{(n-1)!}{(n-r)!(r-1)!}$

To combine terms, we must find a common denominator. Expanding a little reveals the fraction that each term needs to be multiplied by to get the common denominator.

(3) $\quad = \dfrac{(n-1)!}{(n-r-1)!r(r-1)!} + \dfrac{(n-1)!}{(n-r)(n-r-1)!(r-1)!}$

Looking at equation (3), it appears that the left term needs to be multiplied by $\dfrac{n-r}{n-r}$ and the right term by $\dfrac{r}{r}$ in order for the fractions to have a common denominator of $(n-r)(n-r-1)!r(r-1)! = (n-r)!r!$.

(4) $\quad = \dfrac{n-r}{n-r} \times \dfrac{(n-1)!}{(n-r-1)!r(r-1)!} + \dfrac{r}{r} \times \dfrac{(n-1)!}{(n-r)(n-r-1)!(r-1)!}$

(5) $\quad = \dfrac{(n-r)(n-1)!}{(n-r)!r!} + \dfrac{r(n-1)!}{(n-r)!r!}$

(6) $\quad = \dfrac{(n-r)(n-1)!+r(n-1)!}{(n-r)!r!}$

Factoring an (n-1)! from each term in the numerator yields (7)

(7) $\quad = \dfrac{(n-1)!((n-r)+r)!}{(n-r)!r!}$

which simplifies to

(8) $\quad = \dfrac{n!}{(n-r)!r!} = \dbinom{n}{r}.$

This concludes the algebraic argument that $\dbinom{n}{r} =$ $\dbinom{n-1}{r}+\dbinom{n-1}{r-1}.$

Combinatorial Argument: In Pascal's Triangle, Each Term is the Sum of the Two Above It

Similarly, we will example the Combinatorial Argument that

$$\binom{n}{r} = \binom{n-1}{r}+\binom{n-1}{r-1}$$

It just requires a little bit of thinking as to what each of the three terms *mean* and how we can use what they mean to build a convincing argument that the identity is always true. You may want to read this slowly and carefully if you are not at least slightly accustomed to this way of thinking.

First consider the left side of the identity. $\dbinom{n}{r}$ counts the number of ways you can make groups of size r out of n people. The right side of the identity must simply be an alternative way of accomplishing the same task: counting the number of ways you can make groups of size r out of n people.

Here is the alternate way to count the groups. Suppose we have n people and what to make groups of size r. Choose an individual from among all the people. For simplicities sake, let's name him Gauss. Let us consider 2 cases: (1) the number of groups of size r that Gauss is in and (2) the number of groups of size r that Gauss is *not* in. By adding these two cases together, we will have counted all of the groups of size r out of n people: those with Gauss in and those with Gauss out.

Counting the two cases reveals the following. First, count the number of groups of which Gauss is a part. If we put him into the group, then we need r-1 more people. With Gauss already in a group, there are n-1 people remaining to choose from. Consequently, with Gauss included in a group of size r, there are $\binom{n-1}{r-1}$ ways to fill up the group of size r.

The second case is when Gauss is excluded from the group of size r from the start. In this case, we still need r people. However, there are only n-1 to choose from since Gauss is excluded. Consequently, with Gauss excluded from the group of size r, there are $\binom{n-1}{r}$ ways to make groups of size r.

All that must be done is to add the number of groups that exclude Gauss (all $\binom{n-1}{r}$ of them) to the number of groups that include Gauss (the $\binom{n-1}{r-1}$ of them). The sum =

$\binom{n-1}{r} + \binom{n-1}{r-1}$ is then the total number of ways to construct

groups of size r out of n people = $\binom{n}{r}$.

Therefore, it has been shown through a combinatorial argument (an argument that relates to the meaning of the combinations involved) that $\binom{n}{r} = \binom{n-1}{r} + \binom{n-1}{r-1}$.

It is important to note that this proof is more than a combinatorial proof. It actually is the justification that Pascal's Triangle is constructed with combinations. Recall that Pascal's Triangle can generally be simplified as follows: beginning with a generator of 1, each term is the sum of the two above it. Similarly, the exact quote could be used for combinations. First, combinations have a generator of 1 $(\binom{n}{0} = \binom{n}{n} = 1)$. Second, the identity $\binom{n}{r} = \binom{n-1}{r} + \binom{n-1}{r-1}$ shows that each combination is "the sum of the two above it." Hence, Pascal's Triangle = Combinations.

Summary

The two arguments above are examples that contrast algebraic and combinatorial argument. Developing combinatorial argument though patterns apparent in Pascal's Triangle is the goal of this book. It takes a different way of thinking about situations. However, the effort is rewarded with a newly or more highly developed way of thinking. I hope you delight in the journey.

Chapter 4

Patterns in Pascal –

Binomial Probabilities and 2^n

A deeper understanding of combinations will be developed as we explore the various patterns in Pascal's Triangle. The next two patters that will be examined are the occurrence of Binomial Probabilities in Pascal's Triangle as well as the sum of the rows of Pascal's Triangle totaling 2^n in row n.

Recall from chapter 1 that binomial probabilities are those that involve either/or, on/off, up/down dichotomous situations. The simplest version of this may be considering the probabilities involved in flipping any number of coins and finding how many ways one can obtains heads or tails. Dividing that number by the total number of outcomes results in a probability. For instance, if you were to flip 3 coins there are 8 possible outcomes:

HHH – 0 tails

HHT – 1 tails

HTH – 1 tails

HTT – 2 tails

THH – 1 tails

THT – 2 tails

TTH – 2 tails

TTT – 3 tails

Examining the outcomes reveals the third row of Pascal's Triangle:

# of tails	0 tails	1 tails	2 tails	3 tails
# of ways out of 8	1	3	3	1
Probability	1/8	3/8	3/8	1/8

This outcome holds for n coins and r tails:

Number of ways to get r tails → Number (n)of flips ↓	0	1	2	3	4	5	6
0	1						
1	1	1					
2	1	2	1				
3	1	3	3	1			
4	1	4	6	4	1		
5	1	5	10	10	5	1	
6	1	6	15	20	15	6	1

What does this H/T probability result have to do with combinations? Clearly the binomial probability numerators relate to Pascal's Triangle which, as we know from chapter 3, relates to combinations. However, many will not see this as a combination problem immediately. In the first part of this chapter we will justify the numerator of the binomial probabilities while the second part will be devoted to the 2^n denominator of binomial probabilities.

With the proper setting, binomial probabilities will easily be justified combinatorially and Pascal's Triangle will be seen to have a clear relation to binomial probabilities via combinatoric argument. How can we envision flipping coins as combinations? Let us first consider a simple combination problem.

The most common textbook combination problem reads something like this: How many ways can a committee of 3 people be formed from 5 people? The answer is 5 choose 3 or $\binom{5}{3}$. Let us examine the actual committees from the people {A,B,C,D,E}. Recall that order does not matter. Try to list all 10 of them yourself first before reading on.

ABC
ABD
ABE
ACD
ACE
ADE
BCD
BCE
BDE
CDE

This organized list could be looked at based from a slightly different perspective. Namely, the perspective used to prove that the number of ways to choose people to be in the committee is the same as the number of ways to choose people to be off of the committee. Reconsider the 10 results. This time A,B,C,D, and E will always be listed, but the **boldface** letters will be on the committee:

ABC	**ABCDE**
ABD	**ABCDE**
ABE	**ABCDE**
ACD	**ABCDE**
ACE	**ABCDE**
ADE	**ABCDE**
BCD	**ABCDE**
BCE	**ABCDE**
BDE	**ABCDE**
CDE	**ABCDE**

Notice that the committees in this case are formed by assigning committee inclusion as a function of **boldface** being on or off. Can you see where this is going to lead to on/off or H/T binomial probabilities yet?

Consider the flipping of 5 coins in the same sense as choosing committee members: to be on a committee may be "heads" while not being assigned to the committee may be "tails." The chart may be rewritten in many different ways:

Committee By List	Committee By Bold	Committee by "Heads"					5 Coin Flip H/T Summary: Ways to get 3 Heads
		A	B	C	D	E	
ABC	**ABC**DE	H	H	H	T	T	HHHTT
ABD	**AB**C**D**E	H	H	T	H	T	HHTHT
ABE	**AB**CD**E**	H	H	T	T	H	HHTTH
ACD	**A**B**CD**E	H	T	H	H	T	HTHHT
ACE	**A**B**C**D**E**	H	T	H	T	H	HTHTH
ADE	**A**BC**DE**	H	T	T	H	H	HTTHH
BCD	A**BCD**E	T	H	H	H	T	THHHT
BCE	A**BC**D**E**	T	H	H	T	H	THHTH
BDE	A**B**C**DE**	T	H	T	H	H	THTHH
CDE	AB**CDE**	T	T	H	H	H	TTHHH

With this chart it becomes easy to see that the number of ways to make a committee of 3 people from a group of 5 is the same as the number of ways to get 3 heads (or tails) on 5 coin flips: 5 choose 3 or $\binom{5}{3}$. However, it may not be easy to see how this approach is "choosing" anything in particular. An additional approach that clearly involves "choosing" will be helpful as we explore other applications of combinatorial thinking in Pascal's Triangle.

Using the same scenario as above, let us consider the actual 5 coin flips and set this situation up so we are "choosing" things involving flips. There are 5 coins and each one can turn either "heads" or "tails" as indicated by: $(H \quad T)$. Therefore with 5 coins there are 2 options for each coin. We will name the coins A, B, C, D, and E.

A	B	C	D	E
$(H \quad T)$	$(H \quad T)$	$(H \quad T)$	$(H \quad T)$	$(H \quad T)$

The question we will now pursue is the same as we have been: on 5 coin flips, how many different ways can exactly 3 heads appear? Set up in this context, we will choose some coins to turn up heads and other coins to turn up tails. In fact, from the 5 coins we want to know how many ways we can choose 3 to turn up heads. Consequently, among the five parentheses, we want to choose heads from 3 of the sets and tails from 2 of the sets. It should be obvious that this scenario results in the same result as before but involves actually "choosing" something. In this case, we may choose coins ABD or BCE to come up heads. The combinations of coins to come up heads are the same as the list of committees.

The exact value of $\binom{5}{3} = 10$ different ways to get 3 heads on 5 coin flips is not as important as understanding that the choosing of committees in combinations is identical to coin flips and hence binomial probabilities. This is the beginning of combinatorial reasoning: to recognize seemingly unrelated constructs as the simple act of "choosing" combinations. Of course, each time we use combinations we must set the stage for combinations to be obvious. Nevertheless, with this example we can begin to see situations differently: combinatorially.

We can now generalize and speak combinatorially with the above set up in mind. Suppose we flip 5 coins. How many ways can the result be:

(a) exactly 4 heads – answer: $\binom{5}{4}$ (5 parentheses, choose 4 to be heads)

(b) exactly 2 heads – answer: $\binom{5}{2}$ (5 parentheses, choose 2 to be heads)

(c) exactly 1 tails – answer: $\binom{5}{1}$ (5 parentheses, choose 1 to be tails)

(d) at least 2 heads – answer: $\binom{5}{2}+\binom{5}{3}+\binom{5}{4}+\binom{5}{5}$

Each of these calculations is converted to an actual probability once the total number of possible outcomes of the sample space (i.e. the denominator of a probability fraction) is considered.

In summary:

Since:

(1) Pascal's Triangle and combinations are one in the same (see the end of chapter 3)

(2) Binomial probability numerators as examined from the results of any number of H/T coin flips are clearly an example of counting combinations (demonstrated above)

Then:

(3) It should be no surprise that binomial probability numerators appear in Pascal's Triangle. There is nothing mystical about these numbers' appearance; we should be expecting to see binomial probabilities in Pascal's Triangle.

It should be noted that these binomial probability numbers go beyond coin flips. Binomial probabilities are the basis for the "Binomial Distribution" that one finds in statistics which has myriad applications. Binomial probabilities apply to any kind of decision path that contains either/or decisions.

Recall that in Pascal's treatise is divided into two major sections: 19 general observations and 4 specific applications. One of those applications along with much of Pascal's mathematical correspondence relates to games and gambling. Although specific cases involving games of chance and stakes were pursued by earlier mathematicians including Cardano, Tartaglia, Paciolo, and Galileo among others, Pascal and Fermat extensively pursued what called the Problem of Points (Edwards, 144-146). Here is the problem as stated in Edwards extensive modern work *Pascal's Arithmetical Triangle*:

"The Problem of Points, also known as the "division problem", involves determining how the total stake should be equitably divided when a game is terminated prematurely. Suppose two players A and B stake equal money on being the first to win *n* points in a game in which the winner of each point is decided by the toss of a fair coin, heads for A and tails for B. If such a game is interrupted when A still lacks *a* points and B lacks *b*, how should the total stake be divided between them." (146)

It takes little effort for this problem to be pursued once one has the power and understanding of combinations involved in Pascal's Triangle.

The other pattern that should be demystified by this chapter is that the sum of the values in a row of Pascal's Triangle always adds up to a power of 2:

Row	Terms in Series	Sum	Power
0	1	1	2^0
1	1+1	2	2^1
2	1+2+1	4	2^2
3	1+3+3+1	8	2^3
4	1+4+6+4+1	16	2^4
5	1+5+10+10+5+1	32	2^5
n			2^n

Having studied by the binomial probability numerators that are found in Pascal's Triangle, it is a simple corollary that the powers of 2 appear in the triangle. Refer back to the coin tossing scenarios from earlier in the chapter. In row 3 of Pascal's Triangle we find the following binomial probability numerators:

# of flips = 3, r=	0	1	2	3
Number of ways to get r heads	1	3	3	1

The sum of the row is 8 = 2^3. Counting the possibilities was the development of studying the outcomes as choosing an H or T from a set of 2 options:

A	B	C
(H T)	(H T)	(H T)

Notice that for each coin there are 2 options. Therefore there are 2x2x2 = 2^3 possible outcomes. This pattern holds true for all rows. If the values of the rows of Pascal's Triangle represent the number of ways of obtaining exactly 0,1,2,..., n heads when flipping n coins, then there are n sets of 2-option parentheses to choose from: 2 choices, n times equals 2^n possible outcomes. The sums of the rows in Pascal's Triangle *have to be* powers of two. It should be in no way surprising that that pattern is found in Pascal's Triangle. It should be expected.

Chapter 5

Patterns in Pascal –

Binomial Theorem & Powers of 11

In chapter 4 we saw that binomial probabilities ought to be anticipated in Pascal's Triangle in light of a combinatorial understanding of those binomial probabilities. Similarly, the values in the binomial theorem and powers of 11 will be in Pascal's Triangle as obvious consequences of combinatorial thinking.

To review, the binomial theorem relates Pascal's Triangle as follows. A binomial is an algebraic expression with two terms. The simple version of a binomial might be (x+y): two terms added together. Consider the powers of this binomial and their expansions:

Power	Simplified Expansion	Coefficients
$(x+y)^0$	1	1
$(x+y)^1$	x+y	1,1
$(x+y)^2$	$x^2+2xy+y^2$	1,2,1
$(x+y)^3$	$x^3+3x^2y+3xy^2+y^3$	1,3,3,1
$(x+y)^4$	$x^4+4x^3y+6x^2y^2+4xy^3+y^4$	1,4,6,4,1
$(x+y)^5$	$x^5+5x^4y+10x^3y^2+10x^2y^3+5xy^4+y^5$	1,5,10,10,5,1
$(x+y)^6$	$x^6+6x^5y+15x^4y^2+20x^3y^3+15x^2y^4+6xy^5+y^6$	1,6,15,20,15,6,1

The connection to Pascal's Triangle is obvious. More generally, since the values in Pascal's Triangle are combinations, the expansion can be generalized using combinations:

Power	Simplified Expansion	Written with Combinations
$(x+y)^0$	1	$\binom{0}{0}$
$(x+y)^1$	$x+y$	$\binom{1}{0}x+\binom{1}{1}y$
$(x+y)^2$	$x^2+2xy+y^2$	$\binom{2}{0}x^2+\binom{2}{1}xy+\binom{2}{2}y^2$
$(x+y)^3$	$x^3+3x^2y+3xy^2+y^3$	$\binom{3}{0}x^3+\binom{3}{1}x^2y+\binom{3}{2}xy^2+\binom{3}{3}y^3$
$(x+y)^4$	$x^4+4x^3y+6x^2y^2+4xy^3+y^4$	$\binom{4}{0}x^4+\binom{4}{1}4x^3y+\binom{4}{2}x^2y^2+\binom{4}{3}xy^3+\binom{4}{4}y^4$
...		
$(x+y)^n$	$\binom{n}{0}x^n+\binom{n}{1}x^{(n-1)}y^1+\binom{n}{2}x^{(n-2)}y^2+...+\binom{n}{n-2}x^2y^{(n-2)}+\binom{n}{n-1}x^1y^{(n-1)}+\binom{n}{n}y^n$	

The combinatorial thinking involved in understanding this phenomenon is similar to the final explanation of binomial probabilities in chapter 4. First, we must consider what it takes to expand a binomial to a power. What is the actual process? Let's investigate using $(x+y)^4$.

$$(x + y)^4 = (x + y)(x + y)(x + y)(x + y)$$

In order to expand this multiplication, *each* term in *each* binomial must be multiplied to *each* term in every other binomial. Consequently, before simplification by combining like terms, there will be 2x2x2x2=16 terms. For example, one term will be the product of the "x" from the first set a parenthesis, the "y" from the second set, the "y" from the third set, and the "x" from the fourth set resulting in (x)(y)(y)(x) which simplifies to x^2y^2. The question that needs to be asked in terms of coefficients is, "How many x^2y^2 terms out of the 16 terms will there be? This can be examined by exhaustion:

From each of the 4 parenthesis (**bold**)	Term	Simplified	x^2y^2 ??
(**x**+y) (**x**+y) (**x**+y) (**x**+y)	xxxx	x^4	
(**x**+y) (**x**+y) (**x**+y) (x+**y**)	xxxy	x^3y	
(**x**+y) (**x**+y) (x+**y**) (**x**+y)	xxyx	x^3y	
(**x**+y) (**x**+y) (x+**y**) (x+**y**)	xxyy	x^2y^2	①
(**x**+y) (x+**y**) (**x**+y) (**x**+y)	xyxx	x^3y	
(**x**+y) (x+**y**) (**x**+y) (x+**y**)	xyxy	x^2y^2	②
(**x**+y) (x+**y**) (x+**y**) (**x**+y)	xyyx	x^2y^2	③
(**x**+y) (x+**y**) (x+**y**) (x+**y**)	xyyy	xy^3	
(x+**y**) (**x**+y) (**x**+y) (**x**+y)	yxxx	x^3y	
(x+**y**) (**x**+y) (**x**+y) (x+**y**)	yxxy	x^2y^2	④
(x+**y**) (**x**+y) (x+**y**) (**x**+y)	yxyx	x^2y^2	⑤
(x+**y**) (**x**+y) (x+**y**) (x+**y**)	yxyy	xy^3	
(x+**y**) (x+**y**) (**x**+y) (**x**+y)	yyxx	x^2y^2	⑥
(x+**y**) (x+**y**) (**x**+y) (x+**y**)	yyxy	xy^3	
(x+**y**) (x+**y**) (x+**y**) (**x**+y)	yyyx	xy^3	
(x+**y**) (x+**y**) (x+**y**) (x+**y**)	yyyy	y^4	

However, the question as to how many of the 16 are terms that simplify to x^2y^2 could easily be seen as a combinatorial problem. Consider all of the terms that simplified to x^2y^2: in each term y's were chosen from 2 out of the 4 sets of parentheses. The total number of ways to choose 2 y's from 4 sets of parentheses would then be $\binom{4}{2}$. In other words, the coefficient of the x^2y^2 term would be representative of the number of ways of combining 2 y's from the 4 parentheses. In summary, the coefficient of the x^2y^2 term should be $\binom{4}{2}$. There is no other option.

The argument for all coefficients in any binomial expansion of $(x+y)^n$ follow the same line of argument. In the expansion of $(x+y)^7$ we need to determine the coefficient of the x^2y^5 term. When we expand $(x+y)^7$ there are 7 sets of parentheses. From those 7 sets we need to determine the number of ways to choose to multiply 5 of those 7 y's together. The total number of x^2y^5 terms then is 7 parentheses choose 5 y's: $\binom{7}{5} = 21$ ways. Note the symmetry of Pascal's Triangle again. We could have alternately chosen 2 x's from the 7 parentheses for the same result: $\binom{7}{2} = 21$. Similarly, in the expansion of $(x+y)^n$, the x^ny^{n-k} term has a coefficient of $\binom{n}{k} = \binom{n}{n-k}$.

The demystification of the binomial coefficients in Pascal's Triangle should be complete:

1. Pascal's Triangle and Combinations are derived from the same recursive relationship and

2. The way to figure out the coefficient of any term in a binomial expansion of $(x+y)^n$ is through combinations

Therefore,

3. It should be no surprise that Pascal's Triangle can be used as a table to obtain binomial coefficients.

The other pattern discussed in this chapter is the "powers of 11" pattern illustrated in chapter 1. Each row of Pascal's Triangle is a power of 11 when adjusted for our base 10 system. The pattern is obvious in rows 0-4 of Pascal's Triangle but requires the manipulation described in chapter 1 for the subsequent rows.

11^0	1
11^1	11
11^2	121
11^3	1331
11^4	14641

It may at first come as a surprise that these numbers appear but again combinatorial understanding reveals the pattern. In this instance, the argument and result is simply a corollary of the binomial theorem coefficients as described above. Through an argument of choosing x's and y's from sets of parentheses in the simplification of the product $(x+y)^n$, the coefficients result in terms in Pascal's Triangle. The only twist in the "powers of 11" case is this. We need to see 11^3 as $(10+1)^3$. Take a moment to consider how the argument for binomial coefficients applies here. Can you argue that it should be obvious that the powers of 11 will occur in Pascal's Triangle?

Let's consider the multiplication $11^3 = (10 + 1)^3 = (10 + 1)(10 + 1)(10 + 1)$. After multiplying, how do we determine the number in the 1000's place? The 100's place? The 10's place? The 1's place? Consider the analysis as completed above:

Multiplied terms in bold		Multiplied terms extracted	Product
(**10**+1) (**10**+1)	(**10**+1)	10x10x10	1000
(**10**+1) (**10**+1)	(**10**+1)	10x10x1	100
(**10**+1) (10+**1**)	(**10**+1)	10x1x10	100
(**10**+1) (10+**1**)	(10+**1**)	10x1x1	10
(10+**1**) (**10**+1)	(**10**+1)	1x10x10	100
(10+**1**) (**10**+1)	(**10**+1)	1x10x1	10
(10+**1**) (**10**+1)	(10+**1**)	1x1x10	10
(10+**1**) (**10**+1)	(10+**1**)	1x1x1	1

In how many combinations was the product 1000? 100? 10? 1? We may say 1, 3, 3, and 1 respectively. However, it may be more appropriate to say $\binom{3}{0}, \binom{3}{1}, \binom{3}{2},$ and $\binom{3}{3}$ respectively when choosing 1's or not choosing 10's:

Thousands place: 3 parentheses choose 0 1's in the product.

Hundreds place: 3 parentheses choose 1 1's in the product.

Tens place: 3 parentheses choose 2 1's in the product.

Ones place: 3 parentheses choose 3 1's in the product.

It should be clear that we should now expect the powers of 11 in Pascal's Triangle.

Chapter 6

Patterns in Pascal –

Fibonacci

The Fibonacci Sequence itself is a marvel of mathematics. It can be described by a simple recursive formula that starts with 1 and 1 (or some indicate 0 and 1) and then subsequent terms are the sum of the two previous terms. This does sound a bit like Pascal's Triangle and combinations: a generator of 1 where subsequent terms are the sum of the two "above." However, its appearance in Pascal's Triangle is on what seems to be a set of bizarre diagonals.

In review, the Fibonacci Sequence, attributed to Fibonacci (c. 1170 – c. 1250) also known as Leonardo of Pisa (as in leaning tower), is the result of a basic word problem involving rabbits with peculiar reproductive patterns as described in chapter 1. The resulting sequence is 1,1,2,3,5,8,13,21,34,55 ... Each term is the sum of the two previous terms. More formally, the Fibonacci Sequence can be described as $\{f_0 = 1; f_1 = 1; f_n = f_{n-1} + f_{n-2}\}$. Although fairly simple in its construction there are some interesting observations that many have made regarding the Fibonacci Sequence's tie to nature that has raised it to celebrity status in some mathematical and quasi-mathematical circles. The reader is encouraged to google "Fibonacci numbers in nature" or "Fibonacci and the golden section" for an additional extensive excursion into mathematics. However, here is a brief summary.

It is claimed that the Fibonacci numbers occur ubiquitously in nature. From the number of petals on flowers to the number of sunflower seeds and pine cone seeds that run in a sequence around the respective plant, the sequence numbers appear, to the exclusion of other numbers, with eerie regularity. Some claim that this phenomenon is related to the tendency of the Fibonacci sequence toward the so-called "Golden Ratio" or "Golden Section." The Golden Ratio, usually indicated by the Greek letter phi (φ), is a type of mathematically perfect self-repeating ratio numerically equal to $\frac{1+\sqrt{5}}{2}$ that is also claimed to appear in nature with unusually regularity. Some claim that it is a growth ratio of maximum efficiency and hence natural selection highlights its regular appearances. Some theists claim that its appearance is an indication of a universal Designer. I leave the reader to examine the evidence and refutation of so-called "evidence" independently.

What is mathematically certain is that the Fibonacci Sequence and the Golden Ratio are inseparable. There are two major, easily understood connections. First, the ratios of terms

in Fibonacci approach the Golden Ratio.   To illustrate

$\frac{1}{1} = 1, \frac{2}{1} = 2, \frac{3}{2} = 1.5, \frac{5}{3} = 1.\overline{6}, \frac{8}{5} = 1.6$. The ratios zoom in on a

number just greater than 1.6 to exactly φ.   More formally,

$\lim_{n \to \infty} \frac{f_n}{f_{n-1}} = \varphi$.   Secondly, the formula for the Fibonacci

Sequence given above, $\{f_0 = 1; f_1 = 1; f_n = f_{n-1} + f_{n-2}\}$, is

known as a recursive formula because the nth term is based on

values of previous terms.   An explicit formula is one in which

one does not need to know previous terms to calculate the nth

term.   For example, consider the sequence {2, 4, 8, 16, 32,

64,...}.   This sequence can be recursively defined as $\{S_1 =$

$2; S_n = 2 \times S_{n-1}\}$.   However, it can also be easily explicitly

defined as $S_n = 2^n$. Similarly, although the Fibonacci Sequence

can be defined recursively as above, it also has an explicit

formula:[Insert Fib explicit here].   Note the appearance of φ in

the equation.   Hence the second explicit tie between the

Fibonacci Sequence and the Golden Ratio.

Recall from chapter 1 that the Fibonacci Sequence if

found in seemingly rather peculiar way in Pascal's Triangle: as

sums of diagonals as in figure 6.1.

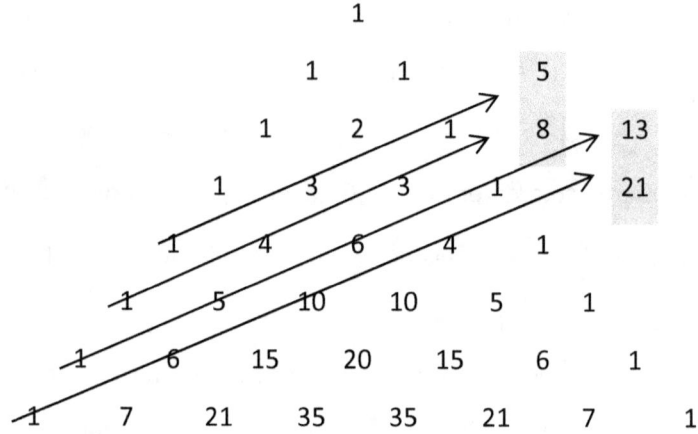

Figure 6.1

Our initial reaction might be that this occurrence of Fibonacci is bizarre but our final conclusion will be that it should be expected if we understand both Fibonacci and combinatorics well.

First, we need to generalize what is happening on these diagonal lines. Examining the terms that sum to Fibonacci Numbers reveals little of intuitive value. However, there are clear patterns that emerge.

f_0	1	1
f_1	1	1
f_2	1+1	2
f_3	1+2	3
f_4	1+3+1	5
f_5	1+4+3	8
f_6	1+5+6+1	13
f_7	1+6+10+4	21
f_8	1+7+15+10+1	34
f_9	1+8+21+20+5	55

Instead, consider the Fibonacci Sequence as it appears in Pascal's Triangle when the triangle is expressed in terms of combinations:

$$\binom{0}{0}$$

$$\binom{1}{0} \quad \binom{1}{1}$$

$$\binom{2}{0} \quad \binom{2}{1} \quad \binom{2}{2}$$

$$\binom{3}{0} \quad \binom{3}{1} \quad \binom{3}{2} \quad \binom{3}{3}$$

$$\binom{4}{0} \quad \binom{4}{1} \quad \binom{4}{2} \quad \binom{4}{3} \quad \binom{4}{4}$$

$$\binom{5}{0} \quad \binom{5}{1} \quad \binom{5}{2} \quad \binom{5}{3} \quad \binom{5}{4} \quad \binom{5}{5}$$

$$\binom{6}{0} \quad \binom{6}{1} \quad \binom{6}{2} \quad \binom{6}{3} \quad \binom{6}{4} \quad \binom{6}{5} \quad \binom{6}{6}$$

$$\binom{7}{0} \quad \binom{7}{1} \quad \binom{7}{2} \quad \binom{7}{3} \quad \binom{7}{4} \quad \binom{7}{5} \quad \binom{7}{6} \quad \binom{7}{7}$$

Examining the terms that sum to the Fibonacci Numbers in this context reveals an easily predictable pattern as well, but one that is simpler. You should be able to predict the next few lines.

f_0	$\binom{0}{0}$	1
f_1	$\binom{1}{0}$	1
f_2	$\binom{2}{0} + \binom{1}{1}$	2
f_3	$\binom{3}{0} + \binom{2}{1}$	3
f_4	$\binom{4}{0} + \binom{3}{1} + \binom{2}{2}$	5
f_5	$\binom{5}{0} + \binom{4}{1} + \binom{3}{2}$	8
f_6	$\binom{6}{0} + \binom{5}{1} + \binom{4}{2} + \binom{3}{3}$	13
f_7	$\binom{7}{0} + \binom{6}{1} + \binom{5}{2} + \binom{4}{3}$	21
f_8	$\binom{8}{0} + \binom{7}{1} + \binom{6}{2} + \binom{5}{3} + \binom{4}{4}$	34
f_9	$\binom{9}{0} + \binom{8}{1} + \binom{7}{2} + \binom{6}{3} + \binom{5}{4}$	55
f_{10}		89
f_{11}		144
f_n	$\binom{n}{0} + \binom{n-1}{1} + \binom{n-2}{2} + \cdots + \binom{\lceil n/2 \rceil}{\lfloor n/2 \rfloor}$; Note: $\lceil n/2 \rceil$ means n/2 rounded up while $\lfloor n/2 \rfloor$ means n/2 rounded down	

What needs to be done now is to develop a combinatorial argument as to why the nth Fibonacci Number

$$f_n = \binom{n}{0} + \binom{n-1}{1} + \binom{n-2}{2} + \cdots + \binom{\lceil n/2 \rceil}{\lfloor n/2 \rfloor}.$$ What do they

have to do with each other?

In order to address the combinatorial argument it will be helpful to examine the behavior of the Fibonacci Sequence from the perspective of two problems. Both of these problems have the same answer: the Fibonacci Sequence. We will examine why the solution to these problems is the Fibonacci Sequence then re-examine the problems from a combinatorial perspective in order to establish the tie between combinations and the Fibonacci Sequence. The result of course be the equation above.

Problem 1: Climbing Stairs. I have seen this problem in various contexts so I do not know which author to attribute it to. However, the problem always goes something like this:

Suppose one can climb a set of stairs by taking steps one at a time or taking steps two at a time or any combination thereof. How many different ways can one climb a set of stairs of length 1? 2? 3? 4? 5?... n?

Let's examine the first few cases. If there is only one step, then there is only 1 possibility: take the one step one at a time. If there are two steps then there are 2 possibilities: take the steps one at a time or take a single "two at a time" step. Now suppose there are three steps. How many ways can the stairs be climbed? 3 possibilities: three single steps, a double followed by a single, and a single followed by a double. Let's denote these three possibilities as (1,1,1), (2,1), and (1,2). Suppose there are four steps. How many ways are possible? (1,1,1,1), (2,1,1), (1,2,1), (1,1,2), (2,2). There are a total of 5 possibilities. Trying to generalize, consider the table:

	Number of Steps	Possibilities	Total Number of Ways
S_1	1	(1)	1
S_2	2	(1,1), (2)	2
S_3	3	(1,1,1), (2,1), (1,2)	3
S_4	4	(1,1,1,1), (2,1,1), (1,2,1), (1,1,2), (2,2)	5
S_5	5	(1,1,1,1,1), (2,1,1,1), (1,2,1,1), (1,1,2,1), (1,1,1,2), (2,2,1), (2,1,2), (1,2,2)	8
S_6	6	(1,1,1,1,1,1), (2,1,1,1,1), (1,2,1,1,1), (1,1,2,1,1), (1,1,1,2,1), (1,1,1,1,2), (2,2,1,1), (2,1,2,1), (2,1,1,2), (1,2,2,1), (1,2,1,2), (1,1,2,2), (2,2,2)	13

The Fibonacci Numbers are conspicuously appearing. However, is there a way we justify or be sure that the result will always be Fibonacci? Why is the answer to the nth term f_n? The Fibonacci recursion can easily be seen. Consider the number of ways of climbing 6 steps. Recontextualize the problems this way: When I take my first step I have two options. My first step can either be a single step or a double step. If my first step is a single step, then I have 5 steps remaining. Therefore, the number of ways to complete my trip with a single step first is the same as the 5 step version of the problem: S_5. However, it I decide to begin my 6-step journey with a double step, then I have 4 steps remaining. Therefore, the number of ways to complete my trip with a double step first is the same as the 4 step version of the problem: S_4. Therefore, $S_6 = S_5 + S_4$. This argument is easily generalized. Hence, given that $S_1 = 1$ and $S_0 = 1$ (zero steps is the trivial case: there is only one way to complete the journey () ← the empty set), we have the following answer to the problem as a recursive function $\{S_0 = 1; S_1 = 1; S_n = S_{n-1} + S_{n-2}\}$ which is identical to Fibonacci: $\{f_0 = 1; f_1 = 1; f_n = f_{n-1} + f_{n-2}\}$.

Problem 2: Domino Covers. This problem can be found in Brualdi's *Introductory Combinatorics* book. The solution again is Fibonacci.

How many ways can you cover a 2 x n checkerboard with 2 x 1 dominos?

Again, let's consider the first few terms then generalize. Visually, here is the problem. We have to cover this:

With these: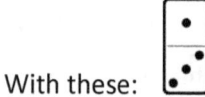

There are two possible ways to produce covers. Any other arrangement will result in 1x1 holes that cannot be covered. We can use a single vertical domino:

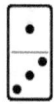

We will denote the use of a vertical domino as 1. The other option is to stack two horizontal dominos:

We will denote the use of two horizontal dominos as 2.

	2x	Covers			Possibilities	
C_1	1				(1)	1
C_2	2				(1,1), (2)	2
C_3	3				(1,1,1), (2,1), (1,2)	3
C_4	4				(1,1,1,1), (2,1,1), (1,2,1), (1,1,2), (2,2)	5

C_5	5		(1,1,1,1,1), (2,1,1,1), (1,2,1,1), (1,1,2,1), (1,1,1,2), (2,2,1), (2,1,2), (1,2,2)	8
C_6	6	I leave this to the reader	(1,1,1,1,1,1), (2,1,1,1,1), (1,2,1,1,1), (1,1,2,1,1), (1,1,1,2,1), (1,1,1,1,2), (2,2,1,1), (2,1,2,1), (2,1,1,2), (1,2,2,1), (1,2,1,2), (1,1,2,2), (2,2,2)	13

After understanding the rationale for the appearance of the Fibonacci Numbers in the "Climbing Stairs" problem, the justification for the appearance of the Fibonacci Numbers in this context should be obvious: When constructing a cover for a 2 x n board you can begin on the left side. You have two options for your first move. You can either begin the cover with a single vertical domino leaving a 2 x (n-1) board to cover or you can begin the cover with two horizontal dominos leaving a 2 x (n-2) board to cover. Assuming the 2 x 0 board has the single trivial empty set cover, we can generalize the following. The total number of covers for a 2 x n checkerboard by 2 x 1 dominos equals the recursive function $\{C_0 = 1; C_1 = 1; C_n = C_{n-1} + C_{n-2}\}$ which is identical to Fibonacci: $\{f_0 = 1; f_1 = 1; f_n = f_{n-1} + f_{n-2}\}$.

The advantage of establishing these two problems as identical to the Fibonacci Sequence allows us to easily build the combinatorial argument for the appearance of the Fibonacci Numbers in Pascal's Triangle. We will first consider a concrete example using n = 6 to establish a single case for the Fibonacci Numbers in Pascal's Triangle. This result will be generalized to the identity $f_n = \binom{n}{0} + \binom{n-1}{1} + \binom{n-2}{2} + \cdots + \binom{\lceil n/2 \rceil}{\lfloor n/2 \rfloor}$.

We will first reconsider the "Climbing Stairs" and "Domino Covers" for the value n = 6. The results in both cases were:

	List of Solutions	Total
$S_6 = C_6$	(1,1,1,1,1,1), (2,1,1,1,1), (1,2,1,1,1), (1,1,2,1,1), (1,1,1,2,1), (1,1,1,1,2), (2,2,1,1,1), (2,1,2,1), (2,1,1,2), (1,2,2,1), (1,2,1,2), (1,1,2,2), (2,2,2)	$13 = f_6$

We can reorganize this list and categorize the solutions:

Number of "double" moves	List of Solutions	Total
0	(1,1,1,1,1,1)	1
1	(2,1,1,1,1), (1,2,1,1,1), (1,1,2,1,1), (1,1,1,2,1), (1,1,1,1,2)	5
2	(2,2,1,1), (2,1,2,1), (2,1,1,2), (1,2,2,1), (1,2,1,2), (1,1,2,2)	6
3	(2,2,2)	1
		1+5+6+1

First notice the right "Total" column should be familiar from this analysis of the Fibonacci Numbers in Pascal's Triangle from above (see **bold**).

f_0	1	1
f_1	1	1
f_2	1+1	2
f_3	1+2	3
f_4	1+3+1	5
f_5	1+4+3	8
f_6	**1+5+6+1**	13
f_7	1+6+10+4	21
f_8	1+7+15+10+1	34
f_9	1+8+21+20+5	55

Now reconsider the chart for n = 6 from a combinatorial perspective. What we are going to be "choosing" is in which position to place the "double" moves. For the case n=6 there are four cases to consider. We can have 0 double moves, 1 double move, 2 double moves, or 3 double moves. Consider the case of 0 double moves. We have 6 slots and want to choose 0 of them to be a double move: $\binom{6}{0} = 1$ combination. Next we consider the case of 1 double move. This is a bit more complicated. If we have to ascend 6 steps and we are going to use 1 double step, then there are four, not five, single steps that must be distributed in our trip. Consequently there are 5 "slots" to fill: 1 double and 4 singles. The question is, "Where do we place the double step?" Given 5 slots and 1 double move, we have $\binom{5}{1} = 5$ options for taking the double move. Alternately, given 5 options {A,B,C,D,E} for a double move, we can choose any of the five positions for the double move. Continuing the argument, we could incorporate 2 double steps and 2 single steps on our ascent. This results in 4 slots and we wish to choose 2 of them to be double steps: $\binom{4}{2} = 6$ combinations. Alternately, given 4 options {A,B,C,D} and I need to choose 2 of these positions for a double move, I can choose A&B or A&C or A&D or B&C or B&D or C&D to be the two double moves. This is precisely what was done when constructing the list of solutions in the table below. The final case is again fairly trivial. When ascending 6 steps we can use 3 double moves resulting in $\binom{3}{3} = 1$ combinations. The table below is reconsidered in light of this argument.

Number of "double" moves	List of Solutions	Total
0	(1,1,1,1,1,1)	$\binom{6}{0} = 1$
1	(2,1,1,1,1), (1,2,1,1,1), (1,1,2,1,1), (1,1,1,2,1), (1,1,1,1,2)	$\binom{5}{1} = 5$
2	(2,2,1,1), (2,1,2,1), (2,1,1,2), (1,2,2,1), (1,2,1,2), (1,1,2,2)	$\binom{4}{2} = 6$
3	(2,2,2)	$\binom{3}{3} = 1$
		$\binom{6}{0} + \binom{5}{1} + \binom{4}{2} + \binom{3}{3}$

This argument is generalized using combinations to lead to the identity:

$$f_n = \binom{n}{0} + \binom{n-1}{1} + \binom{n-2}{2} + \cdots + \binom{\lceil n/2 \rceil}{\lfloor n/2 \rfloor}.$$

Suppose there are n stairs. How many ways can they be climbed using single and double steps? Suppose there is an 2 x n checkerboard. How many ways can it be covered by 2 x 1 dominos?

Number of "double" moves	Number of "slots" to fill	Number of ways to fill the slots with single and double moves
0	n	$\binom{n}{0}$
1	$(n\text{-}1)$	$\binom{n-1}{1}$
2	$(n\text{-}2)$	$\binom{n-2}{2}$
3	$(n\text{-}3)$	$\binom{n-3}{3}$
$\vdots$	$\vdots$	$\vdots$
$\lfloor n/2 \rfloor$	$\lceil n/2 \rceil$	$\binom{\lceil n/2 \rceil}{\lfloor n/2 \rfloor}$
Grand Total:	$\binom{n}{0} + \binom{n-1}{1} + \binom{n-2}{2} + \cdots + \binom{\lceil n/2 \rceil}{\lfloor n/2 \rfloor}$	

Although the Fibonacci Sequence seems at first to appear in a bizarre way in Pascal's Triangle, the following argument changes the appearance of the Fibonacci Sequence from a mystery to a simple anticipated result from a combinatorial perspective:

1. The Fibonacci Sequence and the solutions to both the "Climbing Stairs" and "Domino Covers" problems are identical.

2. There is a simple alternate solution to the "Climbing Stairs" and "Domino Covers" problems that involves combinations.

3. The combinatorial solution and Fibonacci solution to the problems must be identical.

4. Pascal's Triangle and Combinations are derived from the same recursive relationship.

Therefore,

5. It should be no surprise that the Fibonacci Sequence appears in Pascal's Triangle.

Chapter 7

Patterns in Pascal –

Figurate Numbers, Gnomen, & Dimension

As illustrated in chapter 1, figurate numbers have to do with determining the number of stones it takes to make a full geometric figure. Consequently, we have the square numbers (perfect squares as we call them in number theory) 1, 4, 9, 16, 25, 36, ... (Figure 7.1)

1	4	9	16

```
O        O  O      O  O  O      O  O  O  O

         O  O      O  O  O      O  O  O  O

                   O  O  O      O  O  O  O

                                O  O  O  O
```

Figure 7.1

We can also consider pentagonal numbers (numbers of stones to build pentagons), hexagonal numbers, and so on. There are also rectangular numbers. However, triangular numbers are most related to Pascal's Triangle: (Figure 7.2 and 7.3)

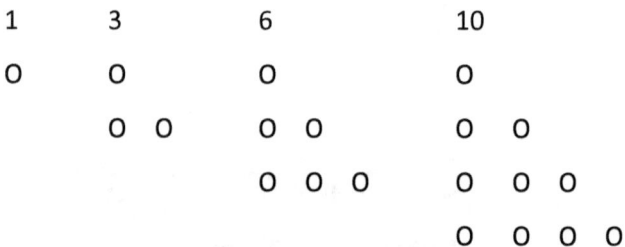

Figure 7.2

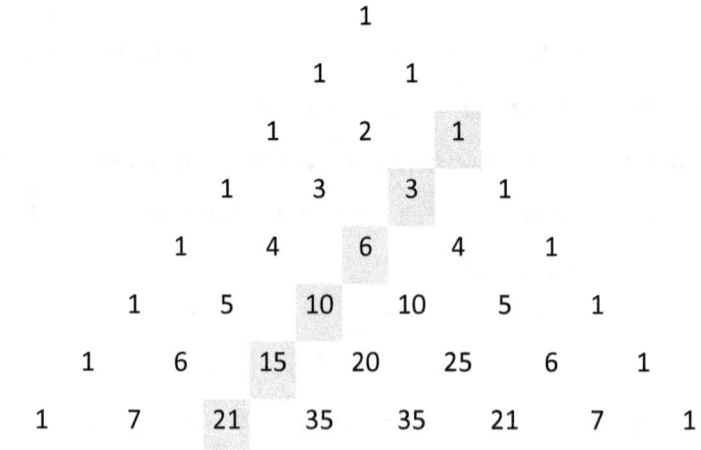

Figure 7.3

All of the patterns in Pascal's Triangle that have been examined to this point continue infinitely through the triangle. Similarly, the figurate numbers are not limited to just this one diagonal; they continue infinitely. However, in order to understand how the pattern deepens and continues, you must think beyond two dimensions.

The easiest way to investigate and extrapolate beyond two dimensions may be to first consider the square numbers in higher dimensions. Consider the square numbers as arranging stones in a square shape in two-dimensions. The analogous figure in three-dimensions is a cube. Think geometrically about the cubic numbers by thinking of building with 6-sided dice as your base objects. Obviously one cube makes a cube. However, it would require eight cubes to make the next size cube (2x2x2). The next cube would require 3x3x3 =27 dice. Four dimensions is considerably more difficult to envision, but easy to extrapolate. Consequently, there are patterns related to the "squares" that continue. Consider the following extrapolations:

Number of dimensions ↓	Points on edge →	1	2	3	4	5	6		N
0 dimensions	Point (no edges)	1	1	1	1	1	1	...	N^0
1 dimension	Segment	1	2	3	4	5	6	...	N^1
2 dimensions	Square	1	4	9	16	25	36	...	N^2
3 dimensions	Cube	1	8	27	64	125	216	...	N^3
4 dimensions	Hyper-cube or Tesseract	1	16	81	256	625	1296	...	N^4
k dimensions		1	2^k	3^k	4^k	5^k	6^k	...	N^k

The square numbers and cubic numbers are not found in Pascal's Triangle. However, examining their pattern is easy and instructive in moving toward an understanding of the triangular numbers and their higher-dimensional analogies. The triangular numbers and their extension are found in Pascal's

Triangle. If one is to extend the pattern, one must consider extending triangular numbers into more dimensions. The three-dimensional analogy of the triangular numbers are the tetrahedral numbers. A tetrahedron is a three dimensional solid that looks like the Egyptian pyramids except the tetrahedron has a triangular base instead of a square base. The easiest way to think about building the tetrahedral numbers is to think in terms of stacking marbles or baseballs instead of using dice. The first tetrahedral number is one as the first triangular number is one. The second tetrahedral number is four as you can construct a tetrahedron with three marbles arranged in a triangle on the "bottom" and one marble on the top. The next two tetrahedral numbers requires ten marbles. (Figure 7.4)

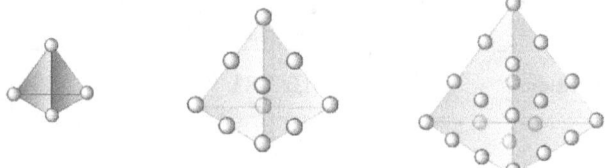

Figure 7.4

The pattern is fairly easily generalized to more dimensions as were the square type numbers of various dimensions:

Number of dim-ensions ↓	Points on edge →	1	2	3	4	5	N
0 dim.	Point (no edges)	1	1	1	1	1	1
1 dim.	Segment	1	2	3	4	5	N
2 dim.	Triangle	1	3	6	10	15	$\dfrac{N(N+1)}{2} = \dfrac{N(N+1)}{2!}$
3 dim.	Tetra-hedron	1	4	10	20	35	$\dfrac{N(N+1)(N+2)}{3!}$
4 dim.	Hyper-tetra-hedron or Penta-tope	1	5	15	35	70	$\dfrac{N(N+1)(N+2)(N+3)}{4!}$
k dim.							$\dfrac{(N+k)!}{(N-1)!\,k!}$

If you cock your head a little you will see Pascal's Triangle emerging. The question remains as to why these numbers appear in Pascal's Triangle.   One simple key to understanding is to develop a deeper understanding of the construction of both figurate numbers and combinations and to recognize their similarities.

The construction of figurate numbers is based on what are called gnomen. Gnomen are essentially the next layer you add to each figurate number. Consider the square number 9. The next layer added to the figure to make the number 16 is referred to as a gnomen:

Square	Gnomen

```
       Square                Gnomen

    O  O  O │ O                   O

    O  O  O │ O                   O

    O  O  O │ O                   O
    _____│
    O  O  O   O           O  O  O  O
```

Examining the gnomen of the square numbers reveals a very interesting pattern:

```
            3        5            7
O  O  O  O       O            O            O

O  O  O  O    O  O            O            O

O  O  O  O             O  O  O            O

O  O  O  O                          O  O  O  O
```

It becomes obvious that the next square number can be found by adding the next consecutive odd number. So the squares are 1, 1+3, 1+3+5, 1+3+5+7, 1+3+5+7+9,... = 1, 4, 9, 16, 25,.... This leads to the obvious recursive formula for the squares: {$SQ_1=1$, $SQ_n = SQ_{n-1} + (2n+1)$} (note: 2n+1 is always an odd number).

Consider the gnomen for the triangular numbers:

```
O

O  O                  → 2

O  O  O               → 3

O  O  O  O            → 4

O  O  O  O  O         → 5
```

So each triangular number can simply be found by taking the previous number and adding a consecutive integer: {$TRI_1=1$; $TRI_n = TRI_{n-1} + n$}. How do the triangular numbers built by gnomen (not built by gnomes) relate to the construction of combinations? There are identical.

The triangular numbers are found in Pascal's Triangle on the third diagonal:

```
                        1
                    1       1
                1       2       1
            1       3       3       1
        1       4       6       4       1
    1       5       10      10      5       1
  1     6       15      20      15      6       1
1     7       21      35      35      21      7       1
```

When Pascal's Triangle is re-expressed in terms of combinations, 1, 3, 6, 10, 15,... are equal to $\binom{2}{2}, \binom{3}{2}, \binom{4}{2}, \binom{5}{2}, \binom{6}{2}, \binom{7}{2}, \binom{8}{2}, \cdots$. Let us carefully examine how $\binom{n}{2}$ is constructed progressively.

We can think of $\binom{n}{2}$ as related to a "handshake" problem; if there are n people in a room and everyone shakes hands with everyone else (but not themselves), how many handshakes occur. Examining this exhaustively, consider the following chart. It examines the handshakes that occur (AB means that person A shakes hands with person B) and begins to illustrate how the triangular numbers relate to combinations.

Number of people	Handshakes	Total
2 (A & B)	AB	1
3 (A, B, & C)	AB AC, BC	3
4 (A, B, C, & D)	AB AC, BC AD, BD, CD	6
5 (A, B, C, D, & E)	AB AC, BC AD, BD, CD AE, BE, CE, DE	10
6 (A, B, C, D, E, & F)	AB AC, BC AD, BD, CD AE, BE, CE, DE AF, BD, CF, DF, EF	15

In this format, one can easily see the gnomen of the triangular numbers. Carefully note the actual elements in each gnomen. When there are 4 people, there are 6 handshakes. When moving to the next row, we can see person E "walking into the room." Since everyone in the room already shook hands, each of A, B, C, and D need only to shake hands with E: 4 more handshakes for a total of 10 handshakes. When the sixth person F walks into the room, 10 handshakes have already occurred and A, B, C, D, and E simply need to complete 5 handshakes with F for a total of 15 handshakes. In other words, to add a fifth person is to add 4 handshakes and to add a sixth person is to add 5 handshakes.

Generalizing, to add a nth person is to add (n-1) handshakes. The construction of both the triangular numbers through gnomen and the constructions of combinations 2 at a time are both dependent on consecutive positive integers: 2,3,4,5.... Combinations taken 2 at a time can be constructed in a way identical to the triangular numbers. It actually provides us with a more geometric understanding of combinations and clearly helps us to understand that triangular numbers should be expected in Pascal's Triangle as it is directly related to combinations.

The argument becomes more difficult when increasing by one dimension to the tetrahedral numbers. However, it is equally as direct. Consider a gnomen approach to building the tetrahedral numbers; each new layer on the bottom of the tetrahedron if formed by a set of marbles arranged in a *triangle*. So each tetrahedral number is simply the previous one plus a triangular number: {$TETRA_1=1$; $TETRA_n= TETRA_{n-1} + TRI_n$}. So we can think of the building of the tetrahedral numbers through gnomen as dependent on the triangular numbers the same way that the construction of the triangular numbers through gnomen as dependent on the consecutive positive integers.

The tetrahedral numbers appear on the fourth diagonal of Pascal's Triangle:

```
                         1

                   1           1

             1           2           1

        1          3           3           1

    1         4           6           4           1

  1        5         10          10          5           1

1        6        15          20          15          6           1

1       7       21         35          35         21          7          1
```

When Pascal's Triangle is written in terms of combinations, the tetrahedral numbers 1, 4, 10, 20, 35,... are equal to

$$\binom{3}{3}, \binom{4}{3}, \binom{5}{3}, \binom{6}{3}, \binom{7}{3}, \binom{8}{3}, \dots.$$

In order to make the combinatorial argument, we will need a three-person analogy to the handshake problem. Let us call a "triumvirate" the three person analogy to the two person handshake. All three people in the triumvirate form a single ring. "ABC" will represent a triumvirate among the three people: A, B, and C. Let us examine the triumvirate problem by exhaustion by building on the handshake problem. The "layers" refer to the top of a tetrahedron, the next layer down, the next layer down, etc.

Number of people	Related Triangular Form	Triumvirates	Total
3 (A, B, & C)	AB	ABC	1
4 (A, B, C, & D)	AB AC, BC	(layer 1) ABC (layer 2) ABD (layer 2) ACD, BCD	4
5 (A, B, C, D, & E)	AB AC, BC AD, BD, CD	(layer 1) ABC (layer 2) ABD (layer 2) ACD, BCD (layer 3) ABE (layer 3) ACE, BCE (layer 3) ADE BDE, CDE	10
6 (A, B, C, D, E, & F)	AB AC, BC AD, BD, CD AE, BE, CE, DE	(layer 1) ABC (layer 2) ABD (layer 2) ACD, BCD (layer 3) ABE (layer 3) ACE, BCE (layer 3) ADE, BDE, CDE (layer 4) ABF (layer 4) ACF, BCF (layer 4) ADF, BDF, CDF (layer 4) AEF, BEF, CEF, DEF	20

Again, take care to notice what happens when a new person "walks into the room." To create all of the new triumvirates, the people have to break back into "handshakes" then each greet the new person in a triumvirate.

The construction of taking *n* items 3 at a time is identical to the gnomen approach to the construction of the tetrahedral numbers. Tetrahedrons again provide a geometric model for understanding combinations taken 3 at a time. The two constructions are identical. Since combinations provide an alternate approach to building Pascal's Triangle, it should be no surprise that the tetrahedral numbers appear.

Continuing the argument to 4 or more dimensions works. However, I leave that construction to the reader. Only think of building 4 dimensional hyper-tetrahedrons by adding gnomen of tetrahedrons (good luck picturing that). Similarly, think of constructing *n* items taken 4 at a time as the creation of "tetra-umvirates or "quad-umvirates" using existing triumvirates. It works.

Chapter 8

Combinatorial Argument –

Pascal's Hockey Stick and

Choices with Replacement

In this final chapter we will practice reasoning combinatorially on a number of additional identities and properties. The first identity we will consider is the one that is produced by examining Pascal's Hockey Stick pattern as identified in chapter 1. We will also develop an argument to justify the fourth "box" in chapter 2 related to combinations: taking n items r at a time when order does NOT matter and there IS replacement, $\binom{n+r-1}{n-1}$. Finally, we will examine additional identities involving combinations and you can try your hand at them.

Pascal's Hockey Stick.

One of the more famous patterns in Pascal's Triangle is the so-called "Hockey Stick" pattern:

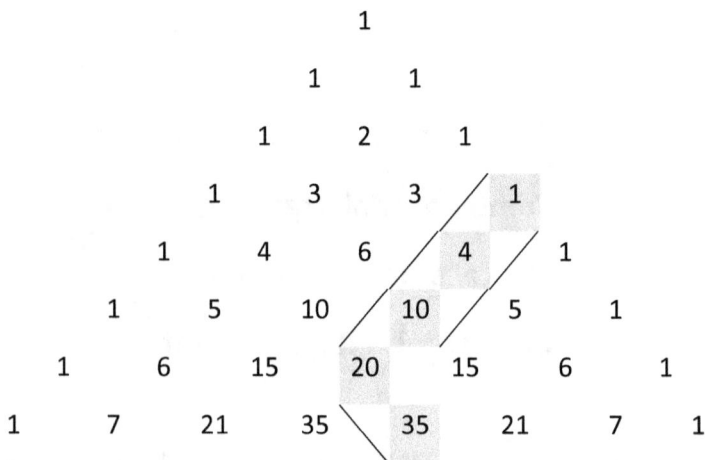

In this pattern, the numbers on the "shaft" of the "hockey stick" add up to the number on the "blade" of the hockey stick.

Before reading on, maybe take a few hours to try to prove this pattern combinatorially.

When Pascal's Triangle is rewritten combinatorially, the hockey stick pattern emerges as a combinatorial identity to be proven:

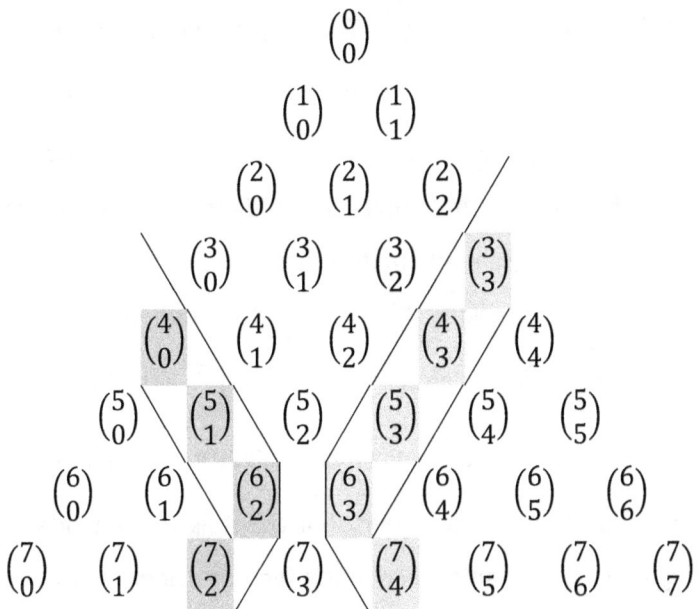

Examine the two of the specific hockey stick patterns highlighted. A left-handed stick version: $\binom{7}{4} = \binom{6}{3} + \binom{5}{3} + \binom{4}{3} + \binom{3}{3}$ and a right-handed stick version: $\binom{7}{2} = \binom{6}{2} + \binom{5}{1} + \binom{4}{0}$. Both types illustrate a possible generalizable identity to prove:

left-handed $\qquad \binom{n}{r} = \binom{n-1}{r-1} + \binom{n-2}{r-1} + \binom{n-3}{r-1} + \cdots + \binom{r-1}{r-1}$ and

right-handed $\qquad \binom{n}{r} = \binom{n-1}{r-1} + \binom{n-2}{r-2} + \binom{n-3}{r-3} + \cdots + \binom{n-r-1}{0}$.

Due to the symmetry of Pascal's Triangle, the identity proven in chapter 3, $\binom{n}{r} = \binom{n}{n-r}$, can be substituted into either equation to change the "handed-ness" of the identity. It would therefore be sufficient to prove either identity. However, we will enjoy both for the mathematical fun of it.

First, we consider the easier left-handed version. In particular, we first examine the case: $\binom{7}{4} = \binom{6}{3} + \binom{5}{3} + \binom{4}{3} + \binom{3}{3}$. In order to argue this equality combinatorially, we need to determine a scenario that shows that counting the left side of the equation is the same as counting the right side. It may be valuable at this point to stop reading for a few hours to attack the problem. Why is the number of ways to make groups of size 4 out of 7 items the exact same thing as making groups of size 3 out of 6 plus groups of size 3 out of 5 plus groups of size 3 out of 4 plus groups of 3 out of 3?

Consider the left side first. The left side determines the number of possible distinct but non-ordered groups of size 4 when there are 7 elements to choose from. The right side of the equation is a construction of the same count. It just makes the list of 7 choose 4 elements in successive steps. For the sake of argument, let the 7 elements be {A, B, C, D, E, F, G}. Then construct 7 choose 4 in steps:

Step 1	Make element A part of every group	Fill the rest of the group with 3 more elements from the remaining 6 elements {B, C, D, E, F, G}	Total number of groups that MUST include element A: $\binom{6}{3}$	
Step 2	Eliminate element A from the possibility of being in a group but guarantee that B will be part of every group	Fill the rest of the group with 3 more elements from the remaining 5 elements {C, D, E, F, G}	Total number of groups that MUST include element B but can NOT include element A: $\binom{5}{3}$	Note: None of the groups on step 2 will be repeats from step 1 because these all do NOT contain A while those in step 1 all contained A
Step 3	Eliminate A and B from the possibility of being in a group but guarantee that C will be part of every group	Fill the rest of the group with 3 more elements from the remaining 4 elements {D, E, F, G}	Total number of groups that MUST include element C but can NOT include elements A or B: $\binom{4}{3}$	Note: None of the groups on step 3 will be repeats from step 1 or 2 because these all do NOT contain A or B while those in step 1 all contained A and those in step 2 all contained B

Step 4	Eliminate A, B, and C from the possibility of being in a group but guarantee that D will be part of every group	Fill the rest of the group with 3 more elements from the remaining 3 elements {E, F, G}	Total number of groups that MUST include element D but can NOT include elements A or B or C: $\binom{3}{3}$	Note: None of the groups on step 3 will be repeats from step 1, 2, or 3 because these all do NOT contain A or B or C while those in step 1 all contained A and those in step 2 all contained B and those in step 3 all contained C

Total Count=	$\binom{6}{3} + \binom{5}{3} + \binom{4}{3} + \binom{3}{3}$

If we understand the above argument combinatorially, then the generalization of the left-handed hockey stick is fairly simple. How do we argue left-handed $\binom{n}{r} = \binom{n-1}{r-1} + \binom{n-2}{r-1} + \binom{n-3}{r-1} + \cdots + \binom{r-1}{r-1}$? The left side of the equation, $\binom{n}{r}$, is a count of the total number of groups of size r when given n elements. For the sake of argument, let the elements be $\{a_1, a_2, a_3, a_4, a_5, \ldots, a_n\}$. Then we can construct the n choose r groups in successive steps. You can provide the argument at this point.

Count the groups that must include a_1, $\binom{n-1}{r-1}$.

Add the groups that exclude a_1 but must include a_2, $\binom{n-2}{r-1}$.

Add the groups that exclude a_1 and a_2 but include a_3, $\binom{n-3}{r-1}$.

Continue this pattern until we construct the single group that uses all $r-1$ of $r-1$ elements in finishing a group. The combinatorial argument is complete.

The right-handed argument uses the same general principles. We must show that $\binom{n}{r} = \binom{n-1}{r-1} + \binom{n-2}{r-2} + \binom{n-3}{r-3} + \cdots + \binom{n-r-1}{0}$. Can you provide the general combinatorial argument? Take some time to think it through. If your get stuck consider this: first, exclude a_1. For the next step include a_1 but exclude a_2. Then include a_1 and a_2 but exclude.... I think you can build the argument.

Combinations with Replacement.

In the general introduction to combinations in chapter 2 we developed the following chart:

Counting Arrangements of n items taken r at a time.	Order matters (AB is different than BA)	Order does NOT matter (AB is the same as BA)
No replacement/ NO repeating of elements	1. Permutations: $nPr = \dfrac{n!}{(n-r)!}$	2. Combinations: $nCr = C(n,r) = \binom{n}{r} = \dfrac{n!}{(n-r)!r!}$
Replacement/ repeating of element	3. n^r	4. $\binom{n+r-1}{n-1}$

At that point, boxes 1, 2, and 3 were developed and justified. However, box 4 was left unjustified until you had developed a better understanding of combinatorial argument. At this point, can you justify the box? It is likely still difficult. However, if we set up the correct scenario, the argument becomes easily accessible.

We will consider an ice-cream parlor with *n flavors*. We want a bowl of ice cream with *r* scoops of ice cream. The criteria are met: There is replacement — flavors can be reused. Order does not matter — we are just placing scoops in a bowl. How many possible bowls of ice cream are there under these conditions? Answer: $\binom{n + r - 1}{n - 1}$. Why? For the sake of argument, we will develop an algorithm for the scooper. The ice cream flavors are in a row (in this order): $[fl_1, fl_2, fl_3, ..., fl_n]$. The scooper works from left-to-right filling the bowl with *r* scoops.

Suppose there are exactly 5 flavors. How many different bowls of ice-cream would there be if a bowl must contain 8 scoops? Using this algorithm, the scooper may do the following:

Scoop flavor 1, Scoop flavor 1, Change to the next flavor, Scoop flavor 2, Change to the next flavor, Scoop flavor 3, Scoop flavor 3, Scoop flavor 3, Change to the next flavor, Scoop flavor 4, Change to the next flavor, Scoop flavor 5. The result is 8 scoops of ice-cream: 2 of flavor 1, 1 of flavor 2, 3 of flavor 3, 1 of flavor 4, and 1 of flavor 5. This algorithm can be used over and over to find all possible bowls. For example, if we let "S" mean scoop and "C" mean change, here is a sample of the possible bowls:

SSSSCSCSCCSS – 8 Scoops: 4 of flavor 1, 1 of flavor 2, 1 of flavor 3, 0 of flavor 4, and 2 of flavor 5.

SCSCSCSCSSSS – 8 Scoops: 1 of flavor 1, 1 of flavor 2, 1 of flavor 3, 1 of flavor 4, and 4 of flavor 5.

SSSSCSSSSCCC – 8 Scoops: 4 of flavor 1, 4 of flavor 2, 0 of flavor 3, 0 of flavor 4, and 0 of flavor 5.

CCCCSSSSSSSS – 8 Scoops: 0 of flavor 1, 0 of flavor 2, 0 of flavor 3, 0 of flavor 4, and 8 of flavor 5.

Now, construct any bowl with 8 scoops from 5 flavors that you wish to have. In every case there are 8 scoops and 4 changes (since there are 5 flavors). Can you anticipate yet how we derive $\binom{n+r-1}{n-1}$?

In the case above there are 5 flavors and 8 scoops. For each example, we had to write 5+8-1 letters for scoops and changes. Furthermore, we had to choose 4 = 5-1 places to change flavors. So there are $\binom{5+8-1}{5-1}$ possible ways to meet the criteria. The general case clearly follows.

Chapter 9

Problems

There are many interesting identities and problems that involve thinking in combinations. This section contains a sampling of those problems. You are encouraged to take time to ponder each question before examining the solution on the following page. Your solution may be more elegant. Try to avoid providing an algebraic argument and try to think combinatorially.

Problem 1

Prove: $\binom{n}{k}\binom{k}{1} = \binom{n}{1}\binom{n-1}{k-1}$

Hint: Consider choosing teams with a captain.

Solution: If we consider the number of ways of choosing a team of size k from among n possible players where one player is the captain of the team, then there are two different ways to build a team. We can choose a team then choose a captain from among the team or choose a captain then build a team around that person. The left and right sides represent these perspectives respectively. $\binom{n}{k}\binom{k}{1}$ is the number of ways to choose a team of size k then choose a captain from among the k players. $\binom{n}{1}\binom{n-1}{k-1}$ is the number of ways to choose a captain from among n players then to build the rest of the k-1 players from among the remaining n-1 options.

Problem 2

How many different ways can people be seated in a classroom if there are 25 people and two rows of 16 chairs where 6 people insist on sitting in the front row, 9 people insist on sitting in the second row, and the other 10 people don't care where they sit?

Hint: Instead of people choosing chairs, consider chairs choosing people.

Solution: First of all, if no one had any preferences then the answer would be a simple $\binom{32}{25}$ options. However, with the restrictions on the problem we must consider satisfying demands one at a time. The 16 chairs in the front must first choose to seat the 6 people. There are $\binom{16}{6}$ ways to assign seats to those 6 people. The 16 chairs in the second row must choose places for the 9 people to sit: $\binom{16}{9}$. For each front row and back row seating, $\binom{16}{6}\binom{16}{9}$, the other 10 people need to be assigned any seat from among the remaining 17 = 32-6-9 seats: $\binom{17}{10}$ options. The total number of seating arrangements would be the product of the three stages as each option must be matched with each other option: $\binom{16}{6}\binom{16}{9}\binom{17}{10}$.

Problem 3

Prove the identity

$$\binom{a}{b}\binom{b}{c} = \binom{a}{b-c}\binom{a-(b-c)}{c}$$

Take some time to try to develop a combinatorial argument for this identity.

Hint: Consider choosing a committee with an internal cabinet or subcommittee.

Solution: Suppose you wanted to choose a committee of size b from among a people. However, within that committee of size b you also want there to be a c member ruling cabinet or subcommittee. This can be done in 2 ways. First, you could choose the committee in $\binom{a}{b}$ ways. Then within the committee you could choose the cabinet of size c from among the b members in a total of $\binom{b}{c}$ ways. The total number of committee structures possible then is $\binom{a}{b}\binom{b}{c}$. Alternately, you could choose the committee in 2 parts: the members *not* on the cabinet then the member on the cabinet. The number of ways to choose the b-c non-cabinet members from among a people is $\binom{a}{b-c}$. Then, the number of ways to choose the c cabinet members from the remaining $a - (b$-$c)$ people (since b-c of the a people were chosen for non-cabinet positions) is $\binom{a - (b - c)}{c}$. The total number of committee structures possible is $\binom{a}{b-c}\binom{a - (b - c)}{c}$. Since both methods count the number of possible committees under the prescribed structure,

$$\binom{a}{b}\binom{b}{c} = \binom{a}{b-c}\binom{a - (b - c)}{c}.$$

Problem 4

$$\text{Prove } \binom{2n}{2} = 2 \times \binom{n}{2} + n^2$$

Hint: Choosing 2 is the same as "handshakes" so $\binom{n}{2}$ is the number of non-repetitive handshakes in a room of n people.

Solution: If $\binom{n}{2}$ is the number of handshakes among n people, then the question is how can we calculate the number of handshakes among 2n people. $\binom{2n}{2}$ is the number of handshakes among 2n people. It can easily be calculated this way. First, break the people up into 2 groups of size n. We'll call them groups A and B. We now consider the number of handshakes *within* groups and the number of handshakes *between* groups. Group A can shake hands among themselves in $\binom{n}{2}$ ways. Similarly, Group B can shake hands among themselves in $\binom{n}{2}$ ways. So just *within* groups there are $2 \times \binom{n}{2}$ total handshakes. Since there are n people in each groups and each of the n people in Group A need to shake hands with each of the n people in Group B, there are $n \times n = n^2$ handshakes *between* groups. The total number of handshakes $\binom{2n}{2}$ equals the total number of *within* groups handshakes, $2 \times \binom{n}{2}$, plus the number of *between* groups handshakes, n^2.

Problem 5

Prove

$$\binom{2n}{n}$$

$$= \binom{n}{0}\binom{n}{n} + \binom{n}{1}\binom{n}{n-1} + \binom{n}{2}\binom{n}{n-2} + \cdots + \binom{n}{n}\binom{n}{0}$$

$$= \binom{n}{0}^2 + \binom{n}{1}^2 + \binom{n}{2}^2 + \cdots + \binom{n}{n}^2$$

(no hint or solution provided)

Selected Bibliography

Benjamin, A. T. & Quinn, J. (2003). *Proofs that really count: The art of combinatorial proof.* MAA.

Burton, D. M. (2003). *The history of mathematics: An introduction* (5th Ed.). Boston: McGraw Hill.

Calinger, R. (Ed.). (1995). *Classics of mathematics.* Upper Saddle River, NJ: Prentice Hall.

Edwards, A. W. F. (2002). *Pascal's arithmetical triangle: The story of a mathematical Idea.* Baltimore: Johns Hopkins.

Katz, V. J. (1993). *A history of mathematics: An introduction.* Harper Collins.

About the Author

Jason VanBilliard is a mathematics education professor who serves as the dean of the School of Liberal Arts and Sciences at Cairn University in Langhorne, Pennsylvania. In addition to teaching undergraduate and graduate courses at Cairn University, he is a regular presenter at local and regional conferences and conducts summer institutes for high school calculus teachers. Jason received his M.A. in Mathematics from West Chester University in West Chester, PA and his Ed.D. in Mathematics Education from Temple University in Philadelphia.